Current Topics in Microbiology

167 and Immunology

Heat Shock Proteins and Immune Response

Edited by S. H. E. Kaufmann

With 18 Figures

Springer-Verlag
Berlin Heidelberg New York
London Paris Tokyo
Hong Kong Barcelona

Professor Dr. STEFAN H. E. KAUFMANN

Dept. of Medical Microbiology and Immunology,
University of Ulm
A.-Einstein-Allee 11, 7900 Ulm, FRG

ISBN-13: 978-3-642-75877-5   e-ISBN-13: 978-3-642-75875-1
DOI:10.1007/978-3-642-75875-1

© Springer-Verlag Berlin Heidelberg 1991
Library of Congress Catalog Card Number 15-12910
Softcover reprint of the hardcover 1st edition 1991

Typesetting: Thomson Press (India) Ltd, New Delhi
Offsetprinting: Saladruck, Berlin; Bookbinding: B. Helm Berlin
2123/3020-543210 – Printed on acid-free paper

Preface

Almost 30 years ago RITOSSA described a new puffing pattern in salivary gland chromosomes of *Drosophila* following heat shock. This was the first description of a heat shock response. For years, development in this field remained modest and it took another decade before the relevant gene products—the heat shock proteins (hsp's)—were made visible by TISSIÈRES and co-workers. Subsequently, progress advanced more rapidly and we can now state that studies on the heat shock response have contributed much to our understanding of various principles in molecular and cellular biology such as control of gene expression and regulation of protein translocation. More recently, the study of hsp's has converged with immunology. There are several reasons for this: The chaperone function of certain hsp's makes them particularly apt for central functions of immunity, including antigen presentation and immunoglobulin synthesis. Furthermore, an effective immune response is often caused or followed by stress situations as they arise during trauma, inflammation, transformation, infection, or autoimmune disease. Due to their abundance during stress, hsp's can provide prominent antigens in many of these situations. This volume contains 11 chapters written by well-known experts dealing with various facets of the fascinating liaison between hsp's and immunity.

The particular relation of hsp's to the immune system may be best illustrated by their intimate association with the major histocompatibility gene complex. Still, as discussed by GÜNTHER, the relevance of this fact to our understanding of hsp functions in immunity remains speculative. The relation of hsp synthesis to fever is also less clear than one might assume at first sight. Yet, evidence from experimental models described by POLLA and KANTENGWA shows that many mediators of inflammation induce hsp synthesis, thus arguing for a central role of hsp's in inflammation.

LANGER and NEUPERT and WELCH and co-workers look at the heat shock response in normal and stressed cells, respectively.

LANGER and NEUPERT focus on the chaperone function of hsp70 and hsp60 in the absence of overt stress, while WELCH et al. concentrate on the response of mammalian cells experiencing metabolic stress situations and on how hsp's facilitate repair and recovery of metabolic pathways in stressed cells.

As described by HAAS, one specialized and well-understood function of hsp70 is its crucial involvement in the assembly of immunoglobulin heavy and light chains. More recent evidence discussed by PIERCE and co-workers indicates that hsp's even participate in the processing of antigens that are recognized by T lymphocytes. The question as to whether the importance of hsp's in host defense against tumor cells is based on such a mechanism of peptide binding and presentation or on the fact that hsp's serve as tumor-associated antigens themselves is discussed by SRIVASTAVA and MAKI.

Hsp's have been found to be major antigens in so many infectious diseases that an in-depth evaluation of their role in the host–parasite relation is urgently warranted. LATHIGRA and co-workers focus on the question as to whether hsp's themselves are virulence factors of microbes, whereas SHINNICK emphasizes the role of hsp's in infectious diseases from the standpoint of the host. Due to their ubiquitous distribution and marked sequence conservation hsp's could provide a link between infection and autoimmune disease. Although it is still unclear why T cells and B cells directed against regions shared by pathogen and host can evade clonal deletion, one can easily envisage how such a situation may result in autoimmunity (WINFIELD and JARJOUR). Perhaps the studies discussed by KAUFMANN and KABELITZ, indicating a particular predilection for hsp's of the newly described T cell population expressing a γ/δ receptor, provide some clue to the role of hsp's in autoimmunity, infection, and transformation.

Studies on hsp's in immunology are at rather an early stage, although enormous progress has been made in the recent past. At least in part, this was due to the convergence of already existing scientific interests. Let us hope that this amalgamation marks the starting point for a productive line of research which will yield deeper insight into both the basic mechanisms of immunology and the more applied aspects of this discipline concerned with the control of infectious, neoplastic, and autoimmune diseases.

STEFAN H. E. KAUFMANN

List of Contents

List of Contributors

(You will find the addresses at the beginning of the respective contribution)

Basic Features of
Heat Shock Proteins

Heat Shock Proteins hsp60 and hsp70:
Their Roles in Folding, Assembly and Membrane Translocation of Proteins

T. LANGER and W. NEUPERT

1 Introduction

Heat shock proteins were initially recognized by their increased expression after exposure of cells to elevated temperatures (RITOSSA 1962; LINDQUIST 1986; LINDQUIST and CRAIG 1988). Subsequently a number of exciting findings stimulated the interest in heat shock proteins. First, homologous proteins were identified in prokaryotic and eukaryotic organisms, suggesting an important general function. Second, it turned out that the various proteins can be grouped into a few distinct families with a high degree of structural conservation during evolution. Third, it was realized that besides elevated temperature other kinds of

Institut für Physiologische Chemie der Universität München, Goethestraße 33, 8000 München 2, FRG

Current Topics in Microbiology and Immunology, Vol. 167
© Springer-Verlag Berlin · Heidelberg 1991

"stress" conditions lead to the induction of proteins which are identical or similar to those induced by heat. For instance, metabolic stress such as glucose starvation or the presence of amino acid analogues was observed to induce a subset of proteins which are immunologically related to heat shock proteins but can be differentiated on the basis of ionic charge (LEE 1987). Fourth, it was realized that some of these heat shock or stress proteins are expressed constitutively in the absence of any kind of stress in various cell types.

Two major lines of research were provoked by these observations. On the one hand, temperature induction proved to be a process well suited to study the molecular basis of regulation of gene expression in both bacteria and eukaryotes. On the other hand, the structure and function of heat shock or stress proteins attracted the attention of researchers in various fields of biology. The present review is not concerned with the complex patterns of regulation of heat shock gene expression (for a review see NEIDHART et al. 1984; LINDQUIST 1986; PELHAM 1989b). It should be pointed out, however, that despite the impressive achievements in defining regulatory elements involved in heat shock regulation, the complete chain of signal transmission from the heat effect to the turning on of specific genes remains elusive. Recent results indicate a role of denatured proteins as a common factor for induction of heat shock proteins (GOFF and GOLDBERG 1985; LEWIS and PELHAM 1985). It is reasonable to assume that the different kinds of stress promote denaturation of intracellular proteins which in turn signal the induction of heat shock proteins (ANANTHAN et al. 1986; KOZUTSUMI et al. 1988). The effect of denatured proteins may be indirect, however, for example by reducing the internal concentration of an unknown regulatory factor.

In the last few years molecular cloning, protein purification, and functional studies in vitro and in vivo have led to a better understanding of the role of heat shock proteins in the sorting, folding, and assembly of proteins, as will be discussed in the present review. It is generally accepted that heat shock proteins protect cells from the damaging effect of temperature or other kinds of metabolic stress. A wide variety of cells have higher thermotolerance after preincubation at an elevated temperature (reviewed in LINDQUIST and CRAIG 1988).

Strong evidence is accumulating that heat shock proteins are necessary for the acquisition of the native structure of monomeric and oligomeric proteins after their synthesis on ribosomes or after transfer across membranes. The transient exposure of hydrophobic or charged residues during these processes can result in misfolding or aggregation of proteins. Therefore, folding in vivo, at least in some cases, seems to require the presence of protein factors. The term "molecular chaperone" was proposed for such proteins (ELLIS 1987), which prevent incorrect interactions and assist assembly without being part of the final structure. It was first used by LASKEY et al. (1978) to illustrate the function of nucleoplasmin in the assembly of nucleosomes. It is now becoming apparent that heat shock proteins can act as chaperones in vitro and in vivo, as will be discussed in detail below. The strongest lines of evidence exist for the members of the hsp60 and hsp70 families. Therefore this review will focus on the role of these proteins in the acquisition of the native structure of proteins in the cell and in the assembly of biological membranes.

2 The hsp70 Family

In each organism examined so far heat shock proteins with a molecular mass of about 70 kD have been found in abundance. They have been observed not only in yeast and mammalian cells, but also in organisms as diverse as plants, trypanosomes, and *Escherichia coli*. hsp70 proteins have a very high degree of evolutionary conservation. Between the bacterial members of the hsp70 family, DnaK, and eukaryotic members, the homology is higher than 45%.

Within the eukaryotic cell hsp70-related proteins were detected in the cytosol compartment, mitochondria (LEUSTEK et al. 1989; MIZZEN et al. 1989; ENGMAN et al. 1989; AMIR-SHAPIRA et al. 1990), chloroplasts (KRISHNASAMY et al. 1989; MARSHALL et al. 1990), and the endoplasmic reticulum (ER). They are members of a multigene family. Some are induced by various stress conditions, others are expressed constitutively. The most thoroughly studied system is that of yeast (as reviewed in detail by LINDQUIST and CRAIG 1988). Nine genes related to the *hsp 70* gene of higher eukaryotes were identified and cloned. Mutational analysis revealed five distinct groups. Ssa1-4, present in the cytosol, are essential for cell viability and may have interchangeable function. hsp70 proteins of the Ssb group (encoded in *ssb1* and *ssb2*), also present in the cytosol, and Ssd1 have not been further characterized. Ssc1 is a mitochondrial protein of 70 627 daltons (CRAIG et al. 1989) with a characteristic amino terminal targeting sequence. Mutant strains in which the *ssc1* gene was deleted were not viable (CRAIG et al. 1987). Recently, *kar 2* was identified as the yeast homologue of the mammalian *BiP/grp78* gene (ROSE et al. 1989; NORMINGTON et al. 1989), which is essential for cell viability and, in addition, is required for nuclear fusion. A cleavable amino terminal hydrophobic signal sequence targets the protein to its site of function, the lumen of the ER.

2.1 General Enzymatic and Structural Properties of hsp70 Proteins

Proteins of the hsp70 family tightly bind ATP, a property that can be used for purification by ATP–agarose chromatography (WELCH and FERAMISCO 1985; CHAPPELL et al. 1986). A weak ATPase activity appears to be a general feature of these proteins. Although several hsp70 proteins have been cloned and sequenced, no consensus sequence for an ATP binding site has been identified so far.

A first structural characterization of hsp70 proteins came from protease degradation studies and mutational analysis. The generation of a stable amino terminal fragment of roughly 44 kD after cleavage with proteases such as chymotrypsin and trypsin suggests a common domain structure of hsp70 proteins. CHAPPELL et al. (1987) purified and characterized this fragment of the uncoating ATPase. They proposed a two-domain model for the hsp70 proteins, which is supported by recent deletion studies with both a human hsp70 (MILARSKI and MORIMOTO 1989) and E. coli DnaK (CEGIELSKA and GEORGOPOULOS 1989). The amino terminal part of the molecules harbors the ATP-binding site and the ATPase activity. CHAPPELL et al. (1987) suggested the term "ATPase core" for this domain. Deletion of the N-terminal region in DnaK and human hsp70 results in the loss of both the ATPase activity and the ATP-binding ability. In the carboxy terminal domain the sequence homology between different hsp70 proteins is relatively low (< 30%). According to the model, this part of the molecule contains activities specific for the different hsp70s, e.g., for the disassembly of coated vesicles in the case of uncoating ATPase. In keeping with the idea of the so-called substrate-binding domain of hsp70 proteins, truncated human hsp70 (lacking carboxy terminal amino acids) shows a lack of proper nucleolar localization after heat shock (MILARSKI and MORIMOTO 1989). A carboxy terminal deletion of DnaK results in the loss of its autophosphorylating activity (CEGIELSKA and GEORGOPOULOS 1989). The so-called substrate-binding domains of other hsp70 molecules remain to be determined.

Although the crystallization of the ATPase fragment of the bovine uncoating ATPase was reported (DELUCA-FLAHERTY et al. 1988), the solution of the structure of an hsp70 protein by X-ray crystallography is still awaited. Preliminary circular dichroism studies of bovine hsc73 and the bacterial DnaK indicate a high content of α-helical structure of both proteins (SADIS et al. 1990).

2.2 Physiological Functions of hsp70 Proteins

The fact that hsp70 proteins were found associated with several cellular and viral proteins in an ATP-dependent manner suggests that direct protein–protein interactions are an essential theme in their function. A first hint at the physiological role of heat shock proteins resulted from determination of their cellular distribution after heat shock (for a review see LINDQUIST and CRAIG 1988). Several heat shock proteins became concentrated in the nucleolar region, including hsp70 proteins (WELCH and SUHAN 1985). They were concentrated in the granular region, the site of preribosomes. Other heat shock proteins, such as hsp110, were found in the fibrillar region, the site of nuclear chromatin (SUBJECK et al. 1983). Nucleoli are heat sensitive and exhibit an altered morphology after heat shock. The recovery of nucleolar morphology occurred more rapidly in the presence of a constitutively expressed hsp70 in the cell, even when synthesis of other heat shock proteins was blocked (PELHAM 1984). This indicates an involvement of the protein in repair processes of RNP structures.

Interaction of hsp70 proteins has been described in the literature with a large number of diverse proteins including: clathrin (UNGEWICKELL 1985; CHAPPELL et al. 1986); the cellular proto-oncogene p53 (PINHASI-KIMHI et al. 1986); certain trans-activating proteins such as E1a of adenovirus (WALTER et al. 1987); malfolded or incompletely assembled proteins in the ER (with BiP, see below); and, as in the case of the prokaryotic hsp70 homologue, DnaK, some proteins involved in phage lambda replication. Recently, immunoprecipitation studies with hsp70-specific antibodies in human cells revealed cell cycle-specific interactions with a number of unidentified proteins (MILARSKI et al. 1989). Interestingly, in these studies coimmunoprecipitation of two different hsp70 proteins was observed. This is consistent with the previously described copurification of these proteins over several steps (WELCH and FERAMISCO 1985) and might suggest a possible functional interaction between these proteins or with identical cellular proteins.

The investigations performed to date show that the hsp70 proteins can be released from their substrates in an ATP-dependent manner in vitro. Nonhydrolyzable ATP analogues have no effect, indicating that ATP hydrolysis is necessary. However, it is not entirely clear whether ATP also has a role in the dissociation of complexes in vivo. Recently, decreased protease sensitivity of BiP was observed after adding adenine nucleotides like ATP and ADP, but not after adding nonhydrolyzable ATP analogues. This may point to a role of the nucleotides in stabilization or induction of different conformations of BiP in the absence of ATP hydrolysis (KASSENBROCK and KELLY 1989). Furthermore, covalent modifications of hsp70 proteins, such as ADP ribosylation, methylation at lysine and arginine residues, or phosphorylation at serine and threonine residues, may be involved in the regulation of their function (WANG and LAZARIDES 1984; HENDERSHOT et al. 1988).

A general model for the action of hsp70 has been proposed that accounts for its ATP-dependent association with other proteins (LEWIS and PELHAM 1985; PELHAM 1986, 1988). Hydrophobic sequences, which are normally buried inside proteins, are exposed after partial denaturation during heat shock or during other cellular processes such as translocation across membranes (see below). This results in an increased tendency for aggregation or misfolding. According to the model, hsp70 proteins prevent or disrupt wrong protein–protein interactions by binding to the exposed regions. This would be reversed with the aid of ATP hydrolysis. Repeated cycles of binding and release could thus repair damage after heat shock or prevent aggregation of proteins. Therefore, the model postulates a general affinity of heat shock proteins for denatured proteins. Although participation of hydrophobic interactions in the binding of hsp70 proteins appears possible, direct evidence for this does not exist.

Recently, an in vitro assay for studying the interaction of hsp70 proteins with model substrates was described (FLYNN et al. 1989). As a first approximation for a native or unfolded protein synthetic peptides were found to bind to purified BiP or hsp70. The ATPase activity of hsp70 and BiP was stimulated by the presence of peptides and caused dissociation of the complexes. Two important observations were made in these studies. First, BiP can bind peptides without added ATP, whereas ATP hydrolysis is necessary for the release. This is consistent with the identification of deletion mutants of human hsp70, which cannot bind ATP but can associate with nucleoli (MILARSKI and MORIMOTO 1989). Second, there was no clear correlation between the overall hydrophobicity of the peptide and its binding affinities. The assay only allows the testing of water-soluble peptides; thus, the affinities for hydrophobic peptides could not be determined. The different affinities of BiP and hsp70 for the various peptides tested may suggest the existence of sites specific for certain sequences or secondary structures rather than for unspecific epitopes. Thus, binding of BiP and hsp70 to peptide segments on the outer surface of a folded protein appears in principle to be possible. This may result in disaggregation or (partial) unfolding of proteins. On the other hand, the free energy of binding could also be used to stabilize conformations which favor certain folding pathways, reduce the tendency for aggregation, or allow membrane translocation (FLYNN et al. 1989).

2.2.1 DnaK—The Prokaryotic Homologue

Among the prokaryotic heat shock proteins identified so far (GEORGOPOULOS et al. 1990), only one, namely DnaK, is homologous to eukaryotic hsp70 protein. The sequence identity to *Drosophila* hsp70 protein is 48% (BARDWELL and CRAIG 1984), and to yeast Ssa1 protein, 49.8% (CRAIG et al. 1989). Mutations in the *dnaK* gene were found to lead to a block in bacteriophage λ DNA replication at all temperatures (GEORGOPOULOS 1977; SUNSHINE et al. 1977). Later on DnaK was characterized as a heat shock protein encoded in the heat shock regulon of *E. coli* (BARDWELL and CRAIG 1984). DnaK is an abundant, consitutively expressed protein with an apparent molecular weight of 70 000. The purified protein (ZYLICZ and GEORGOPOULOS 1984) possesses a weak ATPase activity (with a turnover number of about one ATP per minute) and can be autophosphorylated at threonine residues (ZYLICZ et al. 1983). Its enzymatic activities are well studied (CEGIELSKA and GEORGOPOULOS 1989; DALIE et al. 1990). The ATPase is DNA independent, but is modulated by λO and λP proteins in vitro and in vivo (ZYLICZ et al. 1983). In contrast to ATP binding, ATP hydrolysis and the autophosphorylating activity depend on divalent cations (CEGIELSKA and GEORGOPOULOS 1989; DALIE et al. 1990). Interestingly, Ca^{2+} ions which inhibit the ATPase, stimulate the autophosphorylation activity, indicating a regulatory role of Ca^{2+} (CEGIELSKA and GEORGOPOULOS 1989). The existence of a highly conserved calmodulin-like binding domain in various members of the hsp70 family may be of relevance in this context (STEVENSON and CALDERWOOD 1990).

The role of DnaK in bacteriophage λ replication has been investigated in detail. Besides λO and λP proteins several host proteins are necessary for initiation of DNA replication, including three heat shock proteins, DnaK, DnaJ, and GrpE (reviewed by GEORGOPOULOS et al. 1990). From biochemical and electron microscopic data it became apparent that DnaK participates in an ordered assembly and partial disassembly of the initiation complex, leading to localized DNA unwinding (LIBEREK et al. 1988; ALFANO and McMACKEN 1989a, b; DODSON et al. 1989). In the first step dimeric λO proteins bind specifically to oriλ (about 60 molecules of λO monomers per oriλ; LIBEREK et al. 1988). Complexes of λP protein and the *E. coli* DnaB interact with the resulting nucleosome-like structure. After binding of DnaJ to this prepriming nucleoprotein structure addition of DnaK leads to the complete initiation complex. In the presence of ATP DnaK and DnaJ heat shock proteins cause a partial dissociation of the initiation complex. Thereby, the helicase DnaB is activated in the presence of *E. coli* single strand binding proteins and initiates localized unwinding of the DNA template.

The specific retention of λO and λP proteins on DnaK affinity columns (LIBEREK et al. 1988) strongly suggests an interaction between DnaK and these proteins. The λP proteins were found to bind to the DnaK affin ty column in a salt-resistant manner, suggesting the involvement of hydrophobic interactions, and could be released, at least partially, by ATP hydrolysis. A salt-resistant interaction of DnaK was also observed with another heat shock protein in *E. coli*, the GrpE protein. The binding could be reversed by ATP hydrolysis (ZYLICZ et al. 1987).

In summary, assembly of the initiation complex during phase λ DNA replication requires DnaK. There is also evidence for an invo vement of DnaK in cellular DNA synthesis (SAKAKIBARA 1988). Furthermore, participation of DnaK in reactions other than DNA replication has been reported, e.g., phosphorylation of tRNA synthetases (WADA et al. 1986). The same subset of *E. coli* heat shock proteins seems to be involved in some of these processes, namely DnaK, DnaJ, and GrpE, which are all essential for bacterial growth.

Very recently, first evidence for a role of DnaK in stabilizing precursor proteins destined for secretion was reported (PHILLIPS and SILHAVY 1990). Overproduction of DnaK resulted in increased export of a protein consisting of the signal sequence and the amino terminal region of maltose-binding protein fused to β-galactosidase. A critical role of DnaK in intracellular traffic of precursors was proposed, similar to eukaryotic hsp70 proteins (see below). In addition to DnaK, several other proteins have been suggested to exert such a chaperone function in *E. coli*, namely SecB, trigger factor, and GroEL (COLLIER et al. 1988; CROOKE and WICKNER 1987; CROOKE et al. 1988; BOCHKAREVA et al. 1988).

2.2.2 Catalysis of Clathrin Depolymerization by hsc70

Secreted proteins are transported in specialized coated vesicles from the trans-Golgi to the plasma membrane. The coat consists of the protein clathrin, which forms a latticed cage. Before the vesicle fuses with its target membrane the coat has to dissassemble. The constitutively expressed hsp70, the hsc70, is involved in this process.

Early studies provided evidence for an ATP-dependent enzyme-catalyzed mechanism for clathrin depolymerization (PATZER et al. 1982). The uncoating ATPase was purified based on its ability to release clathrin triskelions from the coat (SCHLOSSMAN et al. 1984). A two-step process was proposed for the uncoating reaction (ROTHMAN and SCHMID 1986). After binding of the uncoating

ATPase to the cage, depending on the presence of ATP and clathrin light chains, ATP hydrolysis would cause a conformational change ("displacement") of a portion of a triskelion, exposing a previously buried site. By binding of the uncoating ATPase to this site, facilitated by ATP or nonhydrolyzable ATP analogues, this conformation would be stabilized ("capture"). After attachment to three points of the triskelion a complex consisting of a clathrin triskelion and three bound enzymes is released.

Immunological cross-reactivity, peptide mapping and two-dimensional gel analysis identified the uncoating ATPase as a constitutively synthesized member of the hsp70 family, called hsc70 (UNGEWICKELL 1985; CHAPPELL et al. 1986).

So far there is no direct evidence that the uncoating reaction is catalyzed by hsc70 in vivo. In view of the high abundance of the uncoating ATPase it is quite possible that disassembly of clathrin coats is not the only function of this heat shock protein. However, the fundamental property of the protein, the ATP-dependent disassembly of protein–protein complexes, would agree with the general view on the function of hsp70 proteins.

2.2.3 Role of Cytosolic hsp70 Proteins in Membrane Translocation of Proteins

A large number of proteins of the eukaryotic cell must be translocated across membranes to reach their functional locations in the various cellular organelles. In many cases this process occurs when polypeptide chain synthesis has been completed (posttranslational translocation). These precursor proteins differ from their native counterparts in several properties: In most cases they contain amino terminal presequences which are cleaved off during or after transit through organelle membranes. Furthermore, precursor proteins usually assume a conformation which is rather different from that of the mature form. A particular requirement for translocation appears to be that precursor proteins are in an unfolded state when traversing the membrane. In a key experiment EILERS and SCHATZ (1986) studied the import into mitochondria of a fusion protein consisting of the presequence of subunit IV of cytochrome oxidase and mouse dihydrofolate reductase (DHFR). Import could be blocked by methotrexate, a substrate analogue, which stabilizes the tertiary structure of DHFR. A similar block of import was obtained when the metallothionein domain was incorporated into a related fusion and a stable tertiary structure was induced by addition of copper (CHEN and DOUGLAS 1987). Conversely, destabilization of tertiary structure by urea or point mutations made import of mitochondrial precursor proteins more efficient. Recent experiments with translocation intermediates spanning both mitochondrial membranes suggest that during import into mitochondria proteins have to undergo extensive unfolding (SCHEYER and NEUPERT 1985; RASSOW et al., unpublished results).

Besides a certain lack of secondary and tertiary structure the presence of ATP seems to be a prerequisite for translocation across membranes. ATP dependency has been found with the import of proteins into the ER, peroxisomes, chloroplasts, mitochondria, and the nucleus. In view of the requirement for ATP and an unfolded conformation of precursor proteins it was proposed several years ago that enzymes (so-called unfoldases) may participate in unfolding using the energy of ATP hydrolysis (ROTHMAN and KORNBERG 1986). On the other hand, a role of hsp70 proteins in folding and assembly of proteins in vivo in an ATP-dependent fashion has been suggested (PELHAM 1986). These proposals have stimulated investigations of a possible role of hsp70 proteins in membrane translocation.

Indeed, recent studies have presented genetic as well as biochemical evidence for a function of hsp70 proteins in posttranslational translocation of proteins across membranes of the ER and mitochondria. By fractionating yeast cytosol on DEAE cellulose, CHIRICO et al. (1988) identified two activities which together stimulate the import of prepro-α-factor into yeast microsomes. One activity, which was insensitive to N-ethylmaleimide (NEM), a sulfhydryl alkylating reagent, was purified using a GTP and an ATP agarose column. It consisted of two members of the yeast hsp70 family, namely Ssa1 and Ssa2. The two proteins are 98% homologous and differ only slightly in their isoelectric point. Whereas ssa1⁻ and ssa2⁻ single mutants lacked any phenotype, double mutants were temperature sensitive for growth, indicating a similar function of the proteins. Ssa1 and Ssa2 proteins had a stimulatory effect on prepro-α-factor import into yeast microsomes in the presence of yeast postribosomal supernatant from the mutant cells. In related experiments import of δ-pyrroline-5-carboxylate-dehydrogenase into yeast mitochondria was found to be stimulated by Ssa1/Ssa2 proteins in the presence of yeast postribosomal supernatant (MURAKAMI et al. 1988). In both studies the activity of the postribosomal supernatant was abolished by NEM treatment, suggesting the involvement of an NEM-sensitive activity besides Ssa1/Ssa2 proteins. Very recently, hsp70 was reported also to stimulate protein import into chloroplast (WAEGEMANN et al. 1990). Thus, a common requirement for hsp70 proteins in protein import into different organelles appears to exist.

The function of hsp70 proteins in the transfer of proteins into the ER was studied further using another heterologous cell free system. The transport of the precursor of M13-phage coat protein (procoat) into dog pancreas microsomes (WIECH et al. 1987) was stimulated by hsc70 (ZIMMERMANN et al. 1988). An increased proteinase K resistance of the procoat protein in the presence of ATP suggested a physical interaction with the heat shock protein.

Studies using yeast mutant strains depleted of hsp70 genes provided valuable additional information (DESHAIES et al. 1988a, b). A strain lacking ssa1, ssa2, and ssa4 genes had been found to be rescued by an ssa1 gene on a single copy plasmid. The ssa1 gene was fused to the yeast gal1 promoter (WERNER-WASHBURNE et al. 1987). Thus, expression could be regulated by growing cells in the presence or absence of galactose. The effect of ssa1 depletion on import of prepro-α-factor into the ER and of $F_1\beta$-ATPase into mitochondria was tested in vivo. After shifting of cells from galactose medium to a glucose medium, prepro-α-factor and $F_1\beta$-ATPase accumulated in the cytosol. Proteolytic processing of the signal sequences did not occur. After partial purification on ATP-agarose the Ssa1 protein stimulated the transport of prepro-α-factor into yeast microsomes.

In conclusion, these observations suggest a general role of the hsp70 proteins in intracellular protein transport. hsp70 proteins may protect proteins from improper folding and interactions and thereby maintain their translocation competence until they reach their final compartment. A role in unfolding of precursor proteins also seems possible although there exists no experimental evidence for such a reaction. The rate of import of urea-denatured precursor proteins in the yeast ER could be increased by adding Ssa1/Ssa2 proteins (CHIRICO et al. 1988). Therefore, hsp70 may slow down refolding of unfolded proteins rather than catalyze unfolding. The localization of hsp70 proteins in different cellular compartments may suggest that they affect precursor proteins in a similar fashion on both sides of an organelle membrane. For further understanding of the functions of hsp70 proteins direct studies of their interaction with precursor proteins will be necessary.

2.2.4 BiP—The hsp70 Homologue in the Endoplasmic Reticulum

The ER contains an hsp70 homologue, which was initially discovered as a protein bound to unassembled immunoglobulin heavy chains; it was therefore called immunoglobulin heavy chain *binding protein* (BiP; HAAS and WABL 1983). Cloning and DNA sequencing of the *BiP* genes of a variety of mammals revealed a high degree of evolutionary conservation (> 98% amino acid identity) and a close relationship (about 60% amino acid identity) to cytoplasmic 70-kD heat shock proteins (MUNRO and PELHAM 1986). The intracellular location of BiP is determined by two signal sequences. A cleavable, hydrophobic signal sequence directs the protein to the ER, whereas the carboxy terminal tetrapeptide KDEL is believed to be responsible for retention of the protein in the lumen of the ER (MUNRO and PELHAM 1987).

BiP differs from other hsp70 proteins with regard to some important features. As shown by sequence analysis, BiP is identical to a 72-kD glucose-regulated protein (initially called grp78), which is not induced by heat (MUNRO and PELHAM 1986; LEE 1987). In contrast to the mammalian BiP, the recently identified yeast homologue (kar2) is induced seven fold by heat (ROSE et al. 1989; NORMINGTON et al. 1989). The rate of BiP synthesis increased after glucose starvation and in the presence of a variety of other substances, including tunicamycin, glucosamine, 2-desoxyglucose, amino acid analogues, and Ca^{2+} ionophores (LEE 1987). The different stress conditions may result in accumulation of malfolded proteins in the ER which have been found to increase the rate of synthesis of glucose-regulated proteins, including BiP (KOZUTSUMI et al. 1988). From these studies it seems likely that malfolding rather than underglycosylation (CHANG et al. 1987) is the primary signal for the induction of BiP, since not all inhibitors of N-glycosylation tested affected the rate of BiP synthesis.

The signal cascade leading to increased BiP synthesis is only partly understood. Posttranslational modifications, including phosphorylation of serine and threonine residues and ADP ribosylation, were suggested to play a role in regulating the synthesis (HENDERSHOT et al. 1988). Conditions leading to increased synthesis of BiP resulted in a decrease in posttranslational modifications. On the other hand, binding of BiP to cellular proteins may be influenced by these modifications (HENDERSHOT et al. 1988). Modified and unmodified BiP coexist in the same compartment, but no modification of BiP molecules associated with other proteins was detected. A possible conclusion is that only the unmodified BiP is responsible for the stress response. In agreement with this, after inhibition of N-glycosylation in a mouse hepatoma cell line only the non-ADP-ribosylated form accumulated (LENO and LEDFORD 1989).

As shown by MUNRO and PELHAM (1986), BiP binds immunoglobulin heavy chains in pre-B cells in an ATP-reversible manner. Binding of BiP to various other proteins was also reversed by ATP hydrolysis, e.g., to nonglycosylated yeast invertase and prolactin containing incorrect disulfide bonds (KASSENBROCK et al. 1988), malfolded and mutant viral glycoproteins (MACHAMER and ROSE 1988; HURTLEY et al. 1989), and hydrophilic peptides (FLYNN et al. 1989). Apparently, BiP has the potential to interact with unassembled or incorrectly folded proteins. BiP possesses a peptide-dependent ATPase activity (KASSENBROCK and KELLY 1989; FLYNN et al. 1989) characterized by a low turnover number and a high affinity for ATP. The decreased sensitivity of BiP to proteolytic degradation in the presence of ATP or ADP suggests that adenine nucleotides may stabilize special conformations of BiP (KASSENBROCK and KELLY 1989). Notably, the existence of a so far unidentified ATP pool in the ER has to be assumed.

Although the precise function of BiP in the ER is unknown, several possibilities are discussed below which are not mutually exclusive.

1. *Retention of proteins in the ER*: The association of BiP with aberrant polypeptides and unassembled immunoglobulin heavy chains might suggest that BiP is part of a quality control system in the ER (HUFTLEY and HELENIUS 1989) which only allows the secretion of functional proteins. First evidence for the importance of correct folding and assembly of proteins for secretion came from experiments which uncovered a correlation between acquisition of native structure and secretion efficiency (extensively discussed by ROSE and DOMS 1988) best studied in the case of abnormally glycosylated proteins (GETHING et al. 1986; DORNER et al. 1987; MACHAMER and ROSE 1988; GALLAGHER et al. 1988). The role of BiP could be to retain the misfolded proteins. The inhibition of N-glycosylation increased the association with BiP, as shown for immunoglobulin heavy chains (BOLE et al. 1986) and several human serum glycoproteins (DORNER et al. 1987). The extent and stability of BiP association were inversely correlated with secretion efficiency. In a more direct approach, DORNER et al. (1988) showed that in CHO cells, expressing plasminogen activator, reduction of BiP levels by introducing antisense RNA led to an increased secretion of the heterologous glycoprotein.

2. *Assembly of oligomeric proteins in the ER*: The interaction of monomeric heavy chains with BiP prior to their association with light chains (HAAS and WABL 1983; BOLE et al. 1986) may point to a role of BiP as an assembly factor (PELHAM 1989a). Heavy chains associated with BiP remained soluble until light chains were expressed. Assembled immunoglobulins were secreted. Proteins which oligomerize in the ER without association with BiP may assemble spontaneously or may interact with other factors as suggested for the T cell receptor (BONIFACINO et al. 1988). However, studies with mutant heavy chains argue against a role of BiP in the assembly of immunoglobulins (HENDERSHOT et al. 1987). Heavy chains lacking the c_H1 domain were not found in association with BiP. The detection of some completely assembled immunoglobulins and the increased rate of secretion, both shown for these mutant heavy chains, could mean that BiP does not assist immunoglobulin assembly, but might prevent the secretion of unassembled heavy chains. On the other hand, the retention of unassembled subunits or assembly intermediates is clearly not a general function of BiP since, with the exception of BiP–heavy chain complexes, stable interactions with incompletely oligomerized proteins have not been detected so far (HURTLEY and HELENIUS 1989).

3. *Folding of proteins in the ER*: Another role of BiP was suggested in the maturation of the VSV-G glycoprotein which forms homotrimers in the ER. In contrast to influenza virus hemagglutinin (HURTLEY et al. 1989), BiP was found associated with monomers of the VSV-G protein shortly after synthesis. Dissociation occurred as the subunits underwent folding (HURTLEY and HELENIUS 1989). No interaction with BiP could be detected by the time trimers were formed in the ER. Mutant forms of VSV-G, which do not fold or trimerize correctly (DOMS et al. 1988), aggregated and were found in association with BiP. Consistent with the kinetics of association, BiP may serve as a folding factor of VSV-G rather than an assembly or retention factor (HURTLEY and HELENIUS 1989). However, it is questionable whether a generalization can be made; e.g., BiP association was not detected in the case of nascent prolactin chains in a cell free translocation system (KASSENBROCK et al. 1988). Still, the possibility exists that BiP may assist specific protein folding in the ER.

4. *Translocation of proteins into the ER*: The identification of the yeast *BiP* gene (ROSE et al. 1989; NORMINGTON et al. 1989; NICHOLSON et al. 1990) allows the examination of BiP function using genetic techniques. At nonpermissive temperature a temperature-sensitive BiP mutant failed to import proteins into the ER (VOGEL et al. 1990). In the absence of BiP function imported precursor proteins might remain bound to a component of the secretory machinery or might aggregate. Both effects would result in inactivation of the translocation machinery resulting in a block of transport into the ER. On the other hand, with translocation-competent proteoliposomes import of preprolactin was found to occur in the absence of BiP, albeit with a low efficiency (NICCHITTA and BLOBEL 1990). However, as discussed by the authors, BiP binding to a polypeptide during membrane translocation may stabilize a conformation that facilitates import and may therefore increase the transloc-

ation rate. Such a function of BiP would resemble that proposed for cytosolic hsp70 proteins in maintaining a transport-competent conformation of translocated proteins. The essential character of BiP function in translocation might be missed in these in vitro experiments due to the limited efficiency of the reconstituted system.

3 The GroEL/hsp60 Family

All prokaryotic and eukaryotic cells investigated so far contain a heat shock protein with a molecular mass of about 60 kD. The first member of this family, the GroEL protein from *E. coli*, was purified and characterized several years ago (HENDRIX 1979; HOHN et al. 1979). Other members were identified in chloroplasts (BARRACLOUGH and ELLIS 1980; PUSHKIN et al. 1982) and mitochondria (MCMULLIN and HALLBERG 1987, 1988). Sequence analysis shows a considerable conservation between the different proteins (about 46%–54% sequence identity; HEMMINGSEN et al. 1988; READING et al. 1989). This homology can be easily explained in terms of the endosymbiotic origin of mitochondria and chloroplasts. So far, identification of a GroEL/hsp60 homologue in the cytosol has not been reported.

In addition to the sequence similarity and immunological cross-reactivity the different members of the hsp60 family have several properties in common. All are constitutively expressed but can be induced by heat shock. Molecular weights and the isoelectric points are almost identical. A close similarity in quarternary structure is obvious (as discussed in detail below). Finally, all members of this family possess a weak ATPase activity, which may be important for their function (HEMMINGSEN et al. 1988).

In view of the strong conservation a common function of these proteins is conceivable. All members of the hsp60 family are thought to assist proteins in the process of acquiring their native structure (HEMMINGSEN et al. 1988). Similar to hsp70 proteins, hsp60 proteins may act as molecular chaperones. The term "chaperonin" was proposed to define them as a third class of chaperone proteins besides nucleoplasmin and the hsp70 protein family (HEMMINGSEN et al. 1988; ELLIS and HEMMINGSEN 1989; ELLIS et al. 1989).

3.1 The Rubisco Subunit Binding Protein of Chloroplasts

Studies on the synthesis of the oligomeric chloroplast enzyme ribulose-1,5-bisphosphate carboxylase/oxygenase (Rubisco) have revealed the association of newly synthesized large subunits with proteins of about 60 kD, termed Rubisco subunit binding protein (BARRACLOUGH and ELLIS 1980). Both were found in a protein particle with an apparent molecular mass of 720 kD and a sedimentation coefficient of 29S (CANNON et al. 1986; ROY and CANNON 1988). There is strong evidence that Rubisco subunit binding protein, absent from native Rubisco, acts as a molecular chaperone during the assembly of Rubisco.

3.1.1 Ribulose-1,5-bisphosphate Carboxylase/Oxygenase (Rubisco)

Rubisco is the most abundant protein in chloroplasts. In higher plants and in most photosynthetic prokaryotes it is a hexadecamer composed of eight large subunits with a moelcular mass of 52 kD and eight small subunits with a molecular mass of 15 kD (MIZIORKO and LORIMER 1983; GATENBY and ELLIS 1990). large subunits, which contain the catalytic site, are encoded by the chloroplast *rbcL* gene, whereas the small subunits are nuclear encoded, synthesized as larger precursors in the cytosol, and processed after import into the chloroplast stroma (SMITH and ELLIS 1979). The small subunits are necessary for enzymatic activity, but their exact function is unclear so far. In the prokaryote *Rhodospirillum rubrum* a dimeric Rubisco exists which comprises two large subunits only (TABITA and MCFADDEN 1974).

Rubisco catalyzes CO_2 fixation, the rate-limiting step in photosynthesis. In addition, it is the key enzyme of photorespiration which reduces the efficiency of CO_2 fixation. Therefore, major efforts were undertaken to manipulate the enzyme in order to increase the net rate of photosynthesis by recombinant DNA techniques. However, large subunits of the higher plant enzyme, when expressed in *E. coli*, formed insoluble aggregates. Thus, after coexpression of large and small Rubisco subunits from higher plants, only low levels of assembly and enzyme activity were observed (GATENBY et al. 1987). This hampered attempts to analyze enzyme structure and function by site-directed mutagenesis experiments.

3.1.2 Structure and Properties

The Rubisco subunit binding protein is one of the most abundant proteins in the stroma of chloroplasts. It consists of two types of subunits with molecular masses of 61 kD (α) and 60 kD(β). These were initially assumed to occur in a stoichiometry of $\alpha_6\beta_6$ (MUSGROVE et al. 1987); however, more recently the subunits were proposed to be identical to a previously identified 14-meric protein (PUSHKIN et al. 1982; HEMMINGSEN et al. 1988). It has not finally been proven, however, that the binding protein is a hetero-oligomer. The possibility of homo-oligomeric isoforms has not been ruled out. The subunits differ in a number of properties, including antigenicity, peptide pattern obtained after limited proteolysis, isoelectric point, and amino terminal amino acid sequence. In rape subunits α and β have about 50% sequence identity (GATENBY and ELLIS 1990). Both subunits are nuclear encoded and translated as precursors with indistinguishable apparent molecular weight (ELLIS and VAN DER VIES 1988). After import into chloroplasts and proteolytic processing they assemble in the stroma compartment.

The protein possesses ATPase activity (CHAUDHARI et al. 1987). As first reported by BLOOM et al. (1983) addition of Mg-ATP in equimolar concentrations causes reversible dissociation of the binding protein into monomeric subunits (MUSGROVE et al. 1987). The disassembly was highly specific for ATP. Other nucleotides together with equimolar amounts of Mg^{2+} ions had no effect. The dissociated subunits were neither stably phosphorylated nor adenylated (HEMMINGSEN and ELLIS 1986).

Immunological studies with an antibody raised against the pea binding protein led to the identification of related proteins in extracts of spinach, tobacco, wheat, and barley leaf extracts and castor bean endosperm. The occurrence of the Rubisco subunit binding protein correlates with the distribution of Rubisco in different plant tissues (HEMMINGSEN and ELLIS 1986; ELLIS and VAN DER VIES 1988).

3.1.3 The Role of Rubisco Subunit Binding Protein in the Assembly of Rubisco

The molecular details of the assembly pathway of Rubisco are largely unknown. On the basis of structural and evolutionary considerations dimerization of folded large subunits was proposed to be a common, conserved step in assembly of dimeric and hexadecameric Rubisco (GOLOUBINOFF et al. 1989a). After oligomerization of dimers to an octameric structure eight small subunits associate polarily, thereby forming the active hexadecameric enzyme.

On the basis of kinetic studies of its association with large subunits the hypothesis was advanced that the Rubisco subunit binding protein is required for the correct assembly of Rubisco (BARRACLOUGH and ELLIS 1980; ROY and CANNON 1988, for a review). Addition of antiserum against binding protein to extracts of pea chloroplasts led to inhibition of holoenzyme formation. This indicates that all assembly-competent large subunits transiently associate with the binding protein (CANNON et al. 1986).

However, the mode of action of Rubisco subunit binding protein is not entirely clear so far. It may affect assembly of Rubisco at various stages. Kinetic studies suggest that large subunits interact with binding proteins before assembly with small subunits occurs (GATENBY et al. 1988). Addition of Mg-ATP to stromal extracts of pea chloroplasts resulted in dissociation of the complex between Rubisco large subunit and Rubisco subunit binding protein. On the other hand, an association of Rubisco subunit binding protein was also found with small subunits (GATENBY et al. 1988). This would suggest a role of Rubisco subunit binding protein for folding of small subunits or for holoenzyme formation.

In summary, there is clear evidence for an involvement of the binding protein in the assembly of Rubisco; however, its exact role remains to be determined.

3.2 GroE Proteins

3.2.1 General Properties

The prokaryotic member of the hsp60 family, the GroEL protein, belongs to the most abundant proteins in *E. coli* and several other bacteria. It is encoded in the GroE operon, which is part of the *E. coli* heat shock regulon. When temperature is elevated from 37 °C to 46 °C the expression of the encoded proteins is increased four- to fivefold. The transcript of about 2100 nucleotides contains two open reading frames. Besides the GroEL protein (molecular weight of 52 259, as estimated from the DNA sequence), a second polypeptide, the GroES protein (molecular weight of 10 368), is encoded in this operon (HEMMINGSEN et al. 1988). The apparent molecular weights of the two proteins determined by denaturing gel electrophoresis are about 65 000 and 15 000 respectively. The GroEL protein forms an oligomeric complex which contains 14 monomers arranged in a double-ring with sevenfold rotational symmetry (HENDRIX 1979; HOHN et al. 1979). Upon gel filtration or centrifugation analysis the GroES protein displays a molecular mass of about 80 kD (CHANDRASEKHAR et al. 1986). This suggests an oligomeric structure for this protein, too. Biochemical as well as genetic evidence exists for an interaction of GroEL and GroES proteins. The GroEL protein possesses a weak ATPase activity (HENDRIX 1979) which can be inhibited by GroES (CHANDRASEKHAR et al. 1986). Furthermore, partial cosedimentation of purified GroES protein with GroEL in a glycerol gradient suggests a physical interaction. Interestingly, ATP and $MgCl_2$ are necessary for this interaction (CHANDRASEKHAR et al. 1986). Under similar conditions GroES binds to immobilized GroEL on an affinity matrix. The identification of intergenic suppressors of *groES* mutations mapping in the *groEL* gene strongly support these biochemical data (TILLY and GEORGOPOULOS 1982).

3.2.2 Function

GroE proteins were originally identified as host genes necessary for bacteriophage T4 morphogenesis (GEORGOPOULOS et al. 1972). Besides the inability to propagate bacteriophages some mutant alleles of both *groEL* and *groES* result in a temperature-sensitive growth of the host (WADA and ITIKAWA 1984). Recently, it was shown by a genetic approach that GroEL and GroES proteins are necessary for bacterial growth at all temperatures (FAYET et al. 1989). The actual level of GroE proteins can determine the maximal growth temperature (KUSUKAWA and YURA 1988). This suggests a more general role of the GroE proteins for cell function. Although their exact mode of action is unclear so far, strong evidence has accumulated during the last few years for an involvement of the GroE proteins in the following cellular processes.

1. *Morphogenesis of bacteriophages:* Even before their identification as heat shock proteins the involvement of the GroE proteins in the morphogenesis of bacteriophages was well established. Both proteins are required for head

assembly of the phage λ (for review, see FRIEDMAN et al. 1984) and tail assembly of T5 phages (TILLY and GEORGOPOULOS 1982). In addition, the GroEL protein has been shown to be necessary for T4 head assembly (GEORGOPOULOS et al. 1972). In all cases the GroE proteins seem to act in early steps of morphogenesis. Examination of the structures formed during λ infection of groEL mutants suggests that this protein may be involved in the oligomerization of the phage B protein to a dodecamer. This ring-like structure is located at the vertex of the λ head to which the λ tail becomes attached.

2. *DNA replication*: The identification of extragenic suppressors of a given mutation can be used to identify functional interactions between different proteins. Overexpression of a DNA fragment containing groEL and groES was found to restore the temperature-sensitive phenotype of dnaA mutations (FAYET et al. 1986; JENKINS et al. 1986). The effect could only be observed in the presence of the DnaA protein, thus excluding a bypass mechanism. Therefore, a direct interaction of the proteins has been proposed. The allele specificity of suppression supports this conclusion. A mutation in the rpoA gene of E. coli, which encodes a subunit of the RNA polymerase, can suppress a temperature-sensitive mutation in the groES gene (WADA et al. 1987). Furthermore, GroEL can rescue a temperature-sensitive mutation in a gene for E. coli single strand binding proteins (Ssb), suggesting an interaction between Ssb proteins and GroEL (RUBEN et al. 1988). It is interesting to note in this context that the levels of GroE proteins were observed to increase with shorter generation times (PEDERSEN et al. 1978). In summary, an involvement of GroE proteins in DNA replication appears to be established; however, their specific function(s) remain to be determined.

3. *Translocation of proteins across membranes*: Posttranslational export of proteins from a prokaryotic cell requires the secreted proteins to be present in a conformation that is conducive to translocation (RANDALL and HARDY 1986). Several proteins, including SecB, trigger factor, and GroEL, have been proposed to assist newly synthesized proteins in acquiring or maintaining such a translocation-competent conformation. They may prevent either folding into a stable native structure or malfolding and aggregation.

Evidence for interaction of GroEL with newly synthesized proteins came from photo-cross-linking experiments (BOCHKAREVA et al. 1988). After such cross-linking newly synthesized, plasmid-encoded secretory β-lactamase (and plasmid-encoded chloramphenicol acetyltransferase) sedimented during ultracentrifugation as a 20S particle. Depletion of E. coli extracts from GroEL by affinity chromatography showed that the particle corresponds to GroEL. Denatured but not native myoglobin caused a competitive inhibition of the cross-linking reaction. This observation suggests an interaction of GroEL with unfolded β-lactamase. Only in the presence of GroEL was the export competence of newly synthesized β-lactamase conserved during preincubation. Therefore, the GroEL protein was proposed to stabilize a secretion-competent conformation and exert a chaperone function analogous to SecB and trigger factor (BOCHKAREVA et al. 1988; KUSUKAWA et al. 1989). As shown recently, the overproduction of GroEL in *E. coli* facilitates the export of a hybrid protein consisting of the signal sequence and the amino terminal region of the maltose-binding protein fused to β-galactosidase (PHILLIPS and SILHAVY 1990).

Employing isolated GroEL and precursor proteins a stable and reversible association of GroEL with the precursors of two secreted proteins, proOmpA and prePhoE, was observed (LECKER et al. 1989). Analysis of the complexes by sucrose gradient centrifugation suggested a 1:1 stoichiometry. In contrast, no interaction of GroEL was seen with soluble cytoplasmic proteins or with mature secreted proteins. However, the significance of these in vitro experiments is not entirely clear: in temperature-sensitive GroEL and GroES mutants only the processing of β-lactamase was slowed down; there was no effect on the secretion of other proteins, including proOmpA (KUSUKAWA et al. 1989). Thus, in the bacterial cell a certain substrate specificity for precursor proteins may exist, e.g., SecB and GroE proteins may differ in this respect.

The studies on GroEL binding to precursor proteins were carried out in the absence of ATP (LECKER et al. 1989). ATP is apparently not necessary for association of the precursor proteins with GroEL. The complexes dissociate after adding ATP (BOCHKAREVA et al. 1988). Since the interaction of GroEL and GroES was only observed in the presence of ATP (CHANDRASEKHAR et al. 1986), a role of the GroES protein in the release of GroEL-associated proteins was assumed (KUSUKAWA et al. 1989). Among different GroES mutants analyzed in respect of processing kinetics in vivo only one affected the export of β-lactamase, suggesting that a specific domain is important for the function of GroES (KUSUKAWA et al. 1989).

The nature of the interaction between GroEL and precursor proteins remains unknown. The specific association with unfolded proteins might point to hydrophobic interactions, in analogy to the hsp70 proteins. As emphasized by LECKER et al. (1989), however, the interaction of GroEL with proOmpA, an integral membrane protein, whose sequence lacks long regions of consecutive apolar residues, might mean that regions other than apolar ones may be recognized.

4. *Protein folding and assembly of oligomeric proteins*: Increasing evidence has accumulated that GroE proteins may assist proteins in the acquisition of their native structure, i.e., act as molecular chaperones. After expression of dimeric or hexadecameric prokaryotic Rubisco in *E. coli* it was shown that overexpression of GroE proteins in *E. coli* resulted in an increased assembly of Rubisco (GOLOUBINOFF et al. 1989a). An influence of GroEL on the transcription rate or the protein stability was excluded. Analysis of GroES and GroEL mutants showed that both GroE proteins are necessary.

In subsequent experiments purified components were used to study the role of GroE proteins in the reconstitution of dimeric prokaryotic Rubisco in vitro (GOLOUBINOFF et al. 1989b). After denaturation with either urea or guanidinium chloride and acid inactivation, no spontaneous reactivation could be observed under the conditions of the assay. In contrast, in the presence of GroEL, GroES, and Mg-ATP efficient reconstitution occurred. In a first step GroEL bound unfolded Rubisco large subunits independently of the presence of GroES and Mg-ATP. The urea-denatured as well as the acid-denatured protein, which was shown to contain a secondary structure, was able to form a binary complex with the GroEL tetradecamer. The subsequent dissociation step depended on GroES and Mg-ATP. In the presence of nonhydrolyzable analogues no effective reconstitution was observed, suggesting that the energy of ATP hydrolysis is required. Optimal reconstitution was observed at equimolar concentrations of GroEL and GroES. GroES may mediate the ATP-dependent release of the protein, perhaps by inducing a conformational change. The inhibition of the ATPase activity of GroEL by GroES (CHANDRASEKHAR et al. 1986) may have a regulatory function in these processes.

In conclusion, the reconstitution experiments with dimeric rubisco indicate that GroE proteins are involved in folding and assembly of large subunits. However, a role in folding and/or assembly of small subunits of hexadecameric Rubiscos of higher plants cannot be excluded. An interaction of GroEL with small subunits was in fact suggested by the observation that small subunits copurified with GroEL after expression in E. coli (LANDRY and BARTLETT 1989).

Several observations are consistent with a more general role of GroE proteins in folding and assembly of proteins. First, the homologous proteins, hsp60 in mitochondria and α-subunit of the Rubisco binding protein in chloroplast, could substitute for GroEL in the reconstitution experiments with Rubisco. So far, no GroES homologue has been identified in mitochondria and chloroplasts. In view of the conservation of GroEL structure and function it seems reasonable to assume that a protein with a function analogous to GroES does exist in these organelles. Furthermore, indirect evidence that GroE proteins assist various proteins in the acquisition of their native structure came from genetic experiments. Overexpressed GroE proteins could suppress many, but not all mutations in different genes of the ilv- and his operon of Salmonella typhimurium (VANDYK et al. 1989). In addition, heat-sensitive folding mutants of gene 9 of Salmonella phage P22 could be rescued by GroE proteins at the restrictive temperature. Thus, the formation not only of active enzymes but also of structural proteins appears to depend on GroE proteins. Interestingly, among the proteins analyzed only mutations in oligomeric ones could be suppressed by GroE proteins; the tested monomeric ones remained unaffected. Although several explanations for the suppressive effect are possible, it seems conceivable that GroE proteins affect the assembly of various proteins.

3.3 The Mitochondrial hsp60

The heat-shock protein hsp60 was initially found in mitochondria of Tetrahymena (MCMULLIN and HALLBERG 1987), and then in mitochondria from all organisms analyzed so far. It has a strong structural similarity to the bacterial

GroEL. Recent studies indicate that it plays an essential role in the folding and assembly of proteins newly imported into mitochondria.

3.3.1 Occurrence and Conserved Properties

Heat shock proteins of the hsp60 type were identified in the yeast *Saccharomyces cerevisiae* (64 kD), *Neurospora crassa* (60 kD; HUTCHINSON et al. 1989), *Tetrahymena thermophila* (MCMULLIN and HALLBERG 1987), *Xenopus laevis* (60 kD), *Zea mays* (62 kD), human cells (58–60 kD; MCMULLIN and HALLBERG 1988; WALDINGER et al. 1988, 1989; JINDAL et al. 1989; MIZZEN et al. 1989; MIZZEN et al. 1989), and CHO cells (58 kD; PICKETTS et al. 1989). Interestingly, the hsp60 of human lymphocytes shows a genetic polymorphism (WALDINGER et al. 1988).

The nuclear-encoded hsp60 protein is constitutively expressed and targeted to mitochondria by a positively charged amino terminal presequence. After heat shock, which causes a two- to threefold increase in expression, the protein represents about 0.3 % of total cell protein. Like the bacterial GroEL homologue and the Rubisco subunit binding protein, the mitochondrial hsp60 assembles into an oligomeric structure in the cell. The protein of *Tetrahymena thermophila* sedimented in sucrose gradients as a 20–25S complex (MCMULLIN and HALLBERG 1987). Electron microscopic analysis of the *Neurospora crassa* protein revealed a structure very similar to that of the GroEL protein. The particle consists of two rings, each comprising seven subunits which are arranged in two layers (HUTCHINSON et al. 1989). The monomeric subunit may have an extended conformation. So far, all available evidence suggests that the 14 subunits are identical; thus, mitochondrial hsp60 appears to be a homo-oligomer like GroEL.

3.3.2 Function

Mitochondrial hsp60 resembles bacterial GroEL with respect to not only supramolecular structure but also amino acid sequence (READING et al. 1989; JOHNSON et al. 1989). Not very surprisingly they appear to be rather similar in function. There is increasing evidence from genetic as well as biochemical studies for an involvement of hsp60 in folding and/or assembly of mitochondrial proteins after their import into the matrix space (CHENG et al. 1989; OSTERMANN et al. 1989; HARTL and NEUPERT 1990, for a review).

The identification of a temperature-sensitive *hsp60* mutant in yeast, called *mif4* (for *m*itochondrial *i*mport *f*unction), allowed examination of the role of hsp60 in the import of proteins into mitochondria (CHENG et al. 1989). At nonpermissive temperature the mutant hsp60 protein became aggregated and was found in the low spin pellet of cell extracts. Under these conditions mitochondrial ornithine transcarbamylase (from humans, transformed into the yeast cells) failed to form trimers and to acquire enzyme activity and the β-subunit of the mitochondrial F_0F_1 ATPase failed to assemble into F_1 particles. Cytochrome b_2, a protein of the mitochondrial intermembrane space, and the Rieske Fe/S protein of complex III, normally present on the outer surface of the inner membrane of mitochondria, did not reach their functional location. In experiments with isolated *mif4* mitochondria it was shown that membrane translocation was still possible, but the imported proteins failed to assemble. Therefore, it was suggested that hsp60 assists oligomeric proteins in the acquisition of the native structure.

In subsequent biochemical studies, a physical interaction of hsp60 with proteins freshly imported after urea denaturation into ATP-depleted mitochondria was established (OSTERMANN et al. 1989). Analysis of the folding state of such hsp60-associated proteins revealed that they were in an unfolded conformation. A fusion protein, in which amino acid residues 1–69 of the precursor of *Neurospora* F_0 ATPase subunit 9 were joined to the amino terminus of mouse dihydrofolate reductase (DHFR), was employed to analyze the folding reaction. The state of folding of this protein could be monitored by determining the protease sensitivity of the DHFR domain. Unfolded DHFR was digested by very low concentrations of proteases, whereas folded DHFR was resistant to rather high concentrations of proteases. After import the fusion protein was found to fold in association with hsp60 in an ATP-dependent manner (OSTERMANN et al. 1989). Folding and release of bound DHFR in the presence of ATP was also observed when mitochondria containing the hsp60 precursor protein complex were lysed with mild detergents. Studies with the membrane-permeant alkylating agent, N-ethylmaleimide (NEM), excluded a spontaneous folding of the DHFR after release from hsp60. In extracts of NEM-treated mitochondria, imported DHFR, bound to hsp60 in the absence of ATP, was released after readdition of ATP, but did not fold correctly. It remains to be tested whether hsp60 itself is the target of the NEM effect.

Furthermore, these experiments suggested an involvement of additional component(s). After partial purification of the fusion protein–hsp60 complex by gel filtration, readdition of Mg-ATP resulted only in a partial protease resistance of the DHFR domain, indicating an incomplete folding reaction. The DHFR remained bound to hsp60 under these conditions. Apparently, an unidentified component which did not cofractionate with the fusion protein–hsp60 complex during gel filtration was necessary for release.

In summary, these findings argue for a role of hsp60 in the folding of proteins after their import into mitochondria. hsp60 was found to be able to substitute for GroEL in assisting the reconstitution of Rubisco in in vitro experiments (GOLOUBINOFF et al. 1989b). This underlines the functional similarity of the members of the GroEL family.

4 Conclusions

Heat shock proteins of the hsp70 and hsp60 classes may fulfill multiple functions in the cell. There is increasing evidence that direct protein–protein interaction is a common theme of their action. The tendency of small heat shock proteins to aggregate and the interaction of hsp90 proteins with steroid receptors and tyrosine kinases (LINDQUIST and CRAIG 1988, for a review) suggests that this may also be true for a variety of other heat shock proteins.

Heat shock proteins protect cells from the damaging effect of high temperatures and other kinds of stress. Although the resolution of aggregated proteins by heat shock proteins has not been observed in vivo or in vitro, the action of hsp70 and hsp60 proteins in the absence of stress may provide hints as to the mechanism of protection. Both families act as molecular chaperones assisting protein folding and assembly by reducing the tendency of aggregation during various cellular processes. However, hsp70 and hsp60 proteins may exert their function in different ways.

hsp70 proteins stabilize protein conformations distinct from the stably folded, native structure. These altered conformations appear to .be necessary for targeting to and transportation of polypeptides across membranes, for assembly into oligomeric structures, or for interactions with other proteins. Thus the shielding of previously buried sequences by hsp70 proteins might reduce the tendency of proteins to aggregate, especially at elevated temperatures. The role of hsp70 proteins in protein folding in vivo is open. The sequence-specific recognition of proteins by BiP and hsc70 in vitro might suggest a function in protein folding (FLYNN et al. 1989). However, up to now no direct experimental data suggest that hsp70 proteins may act as "unfoldases" or as "foldases."

More direct evidence exists for the involvement of hsp60 proteins in folding and assembly of proteins. At least in some cases, GroEL-like proteins have been shown to be necessary for the acquisition of the native structure of proteins in the cell. The complex quaternary structure of hsp60 proteins (in contrast to hsp70 proteins) may be helpful in promoting folding and assembly of proteins. Thus, a general role of hsp60 for protein folding in vivo appears possible. A number of other proteins were identified which assist protein folding in vivo (ROTHMAN 1989; FISCHER and SCHMID 1990; for a review), including cis/trans-peptidyl-prolyl-isomerase (LANG et al. 1987) and the protein disulfide isomerase (FREEDMAN 1989). Within the thermodynamic limits chaperonins and these other proteins may assist folding at a kinetic level. A main task may be the prevention of premature folding and aggregation favored both by the high protein concentration in the cell and by high temperatures.

Acknowledgments. Work in the authors' laboratory was supported by the Fonds der chemischen Industrie and by the Deutsche Forschungsgemeinschaft (SFB 184). We thank Drs. R. Lill, J. Ostermann, and N. Pfanner for critically reading the manuscript.

References

Alfano C, McMacken R (1989a) Ordered assembly of nucleoprotein structures at the bacteriophage λ replication origin during the initiation of DNA replication. J Biol Chem 264: 10699–10708

Alfano C, McMacken R (1989b) Heat shock protein-mediated disassembly of nucleoprotein structures is required for the initiation of bacteriophage λ DNA replication. J Biol Chem 264: 10709–10718

Amir-Shapira D, Leustek T, Dalie B, Weissbach H, Brot N (1990) Hsp70-proteins, similar to *Escherichia coli* DnaK, in chloroplasts and mitochondria of *Euglena gracilis*. Proc Natl Acad Sci USA 87: 1749–1752

Ananthan J, Goldberg AL, Voellmy R (1986) Abnormal proteins serve as eukaryotic stress signals and trigger the activation of heat shock genes. Science 232: 522–524

Bardwell J, Craig E (1984) Major heat shock gene of *Drosophila* and the *Escherichia coli* heat-inducible *dnaK* gene are homologous. Proc Natl Acad Sci USA 81: 848–852

Barraclough R, Ellis RJ (1980) Protein synthesis in chloroplasts. IX: Assembly of newly-synthesized large subunits into ribulose bisphosphate carboxylase in isolated intact pea chloroplasts. Biochem Biophys Acta 608: 19–31

Bloom MV, Milos P, Roy H (1983) Light-dependent assembly of ribulose-1,5-bisphosphate carboxylase. Proc Natl Acad Sci USA 80: 1013–1017

Bochkareva ES, Lissin NM, Girshovich AS (1988) Transient association of newly synthesized unfolded proteins with the heat-shock GroEl protein. Nature 336: 254–257

Bole DG, Hendershot LM, Kearney JF (1986) Posttranslational association of immunoglobulin heavy chain binding protein with nascent heavy chains in nonsecreting and secreting hybridomas. J Cell Biol 102: 1558–1566

Bonifacino JS, Lipincott-Schwartz J, Chen C, Antasch D, Samelson LE (1988) Association and dissociation of the murine T cell receptor associated protein (TRAP). J Biol Chem 263: 8965–8971

Cannon S, Wang P, Roy H (1986) Inhibition of ribulose bisphosphate carboxylase assembly by antibody to a binding protein. J Cell Biol 103: 1327–1335

Cegielska A, Georgopoulos C (1989) Functional domains of the *Escherichia coli* dnaK heat shock protein as revealed by mutational analysis. J Biol Chem 264: 21122–21130

Chandrasekhar GN, Tilly K, Woolford C, Hendrix R, Georgopoulos C (1986) Purification and properties of the groES morphogenetic protein of *Escherichia coli*. J Biol Chem 261: 12414–12419

Chang SC, Wooden SK, Nakaki T, Kim YK, Lin AY, Kung L, Attenello JW, Lee AS (1987) Rat gene encoding the 78-kDa glucose-regulated protein GRP78: its regulatory sequences and the effect of protein glycosylation on its expression. Proc Natl Acad Sci USA 84: 680–684

Chappell TG, Welch WJ, Schlossman DM, Palter KB, Schlesinger MJ, Rothman JE (1986) Uncoating ATPase is a member of the 70 kilodalton family of stress proteins. Cell 45: 3–13

Chappell TG, Konforti BB, Schmid S, Rothman JE (1987) The ATPase core of a clathrin uncoating protein. J Biol Chem 262: 746–751

Chaudhari P, Cannon S, Hubbs A, Roy H (1987) Regulation of Rubisco assembly in pea chloroplast extracts. Abstracts 16th Annual UCLA Symposium. J Cell Biochem [Suppl] 11B: 50

Chen WJ, Douglas MG (1987) The role of protein structure in the mitochondrial import pathway. J Biol Chem 262: 15605–15609

Cheng MY, Hartl FU, Martin J, Pollock RA, Kalousek F, Neupert W, Hallberg EM, Hallberg RL, Horwich AL (1989) Mitochondrial heat-shock protein hsp60 is essential for assembly of proteins imported in yeast mitochondria. Nature 337: 620–625

Chirico WJ, Waters GM, Blobel G (1988) 70K heat shock related proteins stimulate protein translocation into microsomes. Nature 332: 805–810

Collier DN, Bankaitis VA, Weiss JB, Bassford PJ (1988) The antifolding activity of SecB promotes the export of the *E. coli* maltose-binding protein. Cell 53: 273–283

Craig EA, Kramer J, Kosic-Smithers J (1987) SSC1, a member of the 70-kDa heat shock protein multigene family of *Saccharomyces cerevisiae*, is essential for growth. Proc Natl Acad Sci USA 84: 4156–4160

Craig EA, Kramer J, Shilling J, Werner-Washburne M, Holmes S, Kosic-Smithers J, Nicolet CM (1989) SSC1, an essential member of the yeast hsp70 multigene family, encodes a mitochondrial protein. Mol Cell Biol 9: 3000–3008

Crooke E, Wickner W (1987) Trigger factor: a soluble protein that folds pro-OmpA into a membrane-assembly-competent form. Proc Natl Acad Sci USA 84: 5216–5220

Crooke E, Guthrie B, Lecker S, Lill R, Wickner W (1988) ProOmpA is stabilized for membrane translocation by either purified *E. coli* trigger factor or canine signal recognition particle. Cell 54: 1003–1011

Dalie BI, Skaleris DA, Köhle K, Weissbach H, Brot N (1990) Interaction of dnaK with ATP: binding, hydrolysis and Ca^{2+}-stimulated autophosphorylation. Biochem Biophys Res Commun 166: 1284–1292

DeLuca-Flaherty C, Flaherty KM, McIntosh LJ, Bahrami B, McKay DB (1988) Crystals of an ATPase fragment of bovine clathrin uncoating ATPase. J Mol Biol 200: 749–750

Deshaies RJ, Koch BD, Werner-Washburne M, Craig EA, Schekman R (1988a) A subfamily of stress proteins facilitates translocation of secretory and mitochondrial precursor polypeptides. Nature 332: 800–805

Deshaies RJ, Koch BD, Schekman R (1988b) The role of stress proteins in membrane biogenesis. Trends Biochem Sci 13: 384–388

Dodson M, McMacken R, Echols H (1989) Specialized nucleoprotein structures at the origin of replication of bacteriophage λ. J Biol Chem 264: 10719–10725

Doms RW, Ruusala A, Machamer C, Helenius J, Helenius A, Rose JK (1988) Differential effects of mutations in three domains on folding, quarternary structure, and intracellular traffic of vesicular stomatitis virus G protein. J Cell Biol 107: 89–99

Dorner AJ, Bole DG, Kaufman RJ (1987) The relationship of N-linked glycosylation and heavy chain-binding protein association with the secretion of glycoproteins. J Cell Biol 105: 2665–2674

Dorner AJ, Krane KG, Kaufman RJ (1988) Reduction of endogenous GRP78 levels improves secretion of a heterologous protein in CHO cells. Mol Cell Biol 8: 4063–4070

Eilers M, Schatz G (1986) Binding of a specific ligand inhibits import of a purified precursor protein into mitochondria. Nature 322: 228–232

Ellis RJ (1987) Proteins as molecular chaperones. Nature 328: 378–379

Ellis RJ, Hemmingsen SM (1989) Molecular chaperones: proteins essential for the biogenesis of some macromolecular structures. Trends Biochem Sci 14: 339–342

Ellis RJ, van der Vies SM (1988) The Rubisco subunit binding protein. Photosyn Res 16: 101–115

Ellis RJ, van der Vies SM, Hemmingsen SM (1989) The molecular chaperone concept. Biochem Soc Symp 55: 145–153

Engman DM, Kirchhoff LV, Donelson JE (1989) Molecular cloning of mtp70, a mitochondrial member of the hsp70 family. Mol Cell Biol 9: 5163–5168

Fayet O, Louran JM, Georgopoulos C (1986) Suppression of the Escherichia coli dnaA46 mutation by amplification of the groES and groEL genes. Mol Gen Genet 202: 435–445

Fayet O, Ziegelhoffer T, Georgopoulos C (1989) The groES and groEL heat shock gene products of Escherichia coli are essential for bacterial growth at all temperatures. J Bacteriol 171: 1379–1385

Fischer G, Schmid FX (1990) The mechanism of protein folding. Implications of in vitro refolding models for de novo protein folding and translocation in the cell. Biochemistry 29: 2206–2212

Flynn GC, Chappell TG, Rothman JE (1989) Peptide binding and release by proteins implicated as catalysts of protein assembly. Science 245: 385–390

Freedman RB (1989) Protein disulfide isomerase: multiple roles in the modification of nascent secretory proteins. Cell 57: 1069–1072

Friedman DI, Olson ER, Tilly K. Georgopoulos C, Herskowitz I, Banuett F (1984) Interactions of bacteriophage and host macromolecules in the growth of bacteriophage lambda. Microbiol Rev 48: 299–325

Gallagher P, Henneberry J, Wilson I, Sambrook J (1988) Addition of carbohydrate site chains at novel sites on influenza virus hemagglutinin can modulate the folding, transport and activity of the molecule. J Cell Biol 107: 2059–2073

Gatenby AA, Ellis RJ (1990) Chaperone function: the assembly of ribulose bisphosphate carboxylase-oxygenase. Ann Rev Cell Biol (in press)

Gatenby AA, van der Vies SM, Rothstein SJ (1987) Co-expression of both the maize large and wheat small subunit genes of ribulose-bisphosphate carboxylase in Escherichia coli. Eur J Biochem 168: 227–231

Gatenby AA, Lubben TH, Ahlquist P, Keegstra K (1988) Imported large subunits of ribulose bisphosphate carboxylase/oxygenase, but not imported β-ATP synthase subunits, are assembled into holoenzyme in isolated chloroplasts. EMBO J 7: 1307–1314

Georgopoulos C (1977) A new bacterial gene (groPC) which affects lambda DNA replication. Mol Gen Genet 151: 35–39

Georgopoulos C, Hendrix R, Kaiser A, Wood W (1972) Role of the host cell in bacteriophage morphogenesis: effects of a bacterial mutation on T4 head assembly. Nature New Biol 239: 38–41

Georgopoulos C, Ang D, Liberek K, Zylicz M (1990) Properties of the Escherichia coli heat shock proteins and their role in bacteriophage λ growth. In: Morimoto R, Tissieres A, Georgopoulos C (eds) Stress proteins in biology and medicine. Cold Spring Harbor Laboratory, Cold Spring Harbor NY

Gething MJ, McCommon K, Sambrook J (1986) Expression of wild-type and mutant forms of influenza hemagglutinin: the role of folding in intracellular transport. Cell 46: 939–950

Goff SA, Goldberg AL (1985) Production of abnormal proteins in *E. coli* stimulates transcription of ion and other heat shock genes. Cell 41: 587–595

Goloubinoff P, Gatenby AA, Lorimer GH (1989a) GroE heat-shock proteins promote assembly of foreign prokaryotic ribulose bisphosphate carboxylase oligomers in *Escherichia coli*. Nature 337: 44–47

Goloubinoff P, Christeller JT, Gatenby AA, Lorimer GH (1989b) Reconstitution of active dimeric ribulose bisphosphate carboxylase from an unfolded state depends on two chaperonin proteins and Mg-ATP. Nature 342: 884–889

Haas IG, Wabl M (1983) Immunoglobulin heavy chain binding protein. Nature 306: 387–389

Hartl FU, Neupert W (1990) Protein sorting to mitochondria: evolutionary conservations of folding and assembly. Science 247: 930–938

Hemmingsen SM, Ellis RJ (1986) Purification and properties of ribulosebisphosphate carboxylase large subunit binding protein. Plant Physiol 80: 269–276

Hemmingsen SM, Woolford C, van der Vies SM, Tilly K, Dennis DT, Georgopoulos CP, Hendrix RW, Ellis RJ (1988) Homologous plant and bacterial proteins chaperone oligomeric protein assembly. Nature 333: 330–334

Hendershot L, Bole D, Köhler G, Kearney JF (1987) Assembly and secretion of heavy chains that do not associate posttranslationally with immunoglobulin heavy chain-binding protein. J Cell Biol 104: 761–767

Hendershot LM, Ting J, Lee AS (1988) Identity of the immunoglobulin heavy-chain-binding protein with the 78,000 glucose-regulated protein and the role of posttranslational modifications in its binding function. Mol Cell Biol 8: 4250–4256

Hendrix RW (1979) Purification and properties of groE, a host protein in bacteriophage assembly. J Mol Biol 129: 375–392

Hohn T, Hohn B, Engel A, Wurtz M (1979) Isolation and characterization of the host protein groE involved in bacteriophage lambda assembly. J Mol Biol 129: 359–373

Hurtley SM, Helenius A (1989) Protein oligomerization in the endoplasmic reticulum. Ann Rev Cell Biol 5: 277–307

Hurtley SM, Bole DG, Hoover-Litty H, Helenius A, Copeland CS (1989) Interactions of misfolded influenza virus hemagglutinin with binding protein (BiP). J Cell Biol 108: 2117–2126

Hutchinson EG, Tichelaar W, Hofhaus G, Weiss H, Leonard KR (1989) Identification and electron microscope analysis of a chaperonin oligomer from *Neurospora crassa* mitochondria. EMBO J 8: 1485–1490

Jenkins AJ, March JB, Oliver IR, Masters M (1986) A DNA fragment containing the *groE* genes can suppress mutations in the *Escherichia coli dnaA* gene. Mol Gen Genet 202: 446–454

Jindal S, Dudani AK, Singh B, Harley AB, Gupta RS (1989) Primary structure of a human mitochondrial protein homologous to the bacterial and plant chaperonins and to the 65-kilodalton mycobacterial antigen. Mol Cell Biol 9: 2279–2283

Johnson RB, Fearon K, Mason T, Jindal S (1989) Cloning and characterization of the yeast chaperonin HSP60 gene. Gene 84: 295–302

Kassenbrock CK, Kelly RB (1989) Interaction of heavy chain binding protein (BiP/GRP78) with adenine nucleotides. EMBO J 8: 1461–1467

Kassenbrock CK, Garcia PD, Walter P, Kelly RB (1988) Heavy-chain binding protein recognizes aberrant polypeptides translocated in vitro. Nature 333: 90–93

Kozutsmi Y, Segal M, Normington K, Gething MJ, Sambrook J (1988) The presence of malfolded proteins in the endoplasmic reticulum signal the induction of glucose-regulated proteins. Nature 332: 462–464

Krishnasamy S, Mannar Mannan R, Krishnan M, Gnanam A (1988) Heat shock response of the chloroplast genome in *Vigna sinensis*. J Biol Chem 11: 5104–5109

Kusukawa N, Yura T (1988) Heat-shock protein groE of *Escherichia coli*—key protective roles against thermal stress. Genes Dev 2: 874–882

Kusukawa N, Yura T, Ueguchi C, Akiyama Y, Ito K (1989) Effects of mutations in heat-shock genes *groEL* on protein export in *Escherichia coli*. EMBO J 8: 3517–3521

Landry SJ, Bartlett SG (1989) The small subunit of ribulose-1,5-bisphosphate carboxylase/oxygenase and its precursor expressed in *Escherichia coli* are associated with GroEL protein. J Biol Chem 264: 9090–9093

Lang K, Schmid FX, Fischer G (1987) Catalysis of protein folding by prolyl isomerase. Nature 329: 268–270

Laskey RA, Honda BM, Mills AD, Finch JT (1978) Nucleosomes are assembled by an acidic protein which binds histones and transfers them to DNA. Nature 275: 416–420

Lecker S, Lill R, Ziegelhoffer T, Georgopoulos C, Bassford PJ, Kumamoto CA, Wickner W (1989) Three pure chaperone proteins of *Escherichia coli*—SecB, trigger factor and GroEL—form soluble complexes with precursor proteins in vitro. EMBO J 8: 2703–2709

Lee AS (1987) Coordinated regulation of a set of genes by glucose and calcium ionophores in mammalian cells. Trends Biochem Sci 12: 20–23

Leno GH, Ledford BE (1989) ADP-ribosylation of the 78-kDa glucose-regulated protein during nutritional stress. Eur J Biochem 186: 205–211

Leustek T, Dalie B, Amir-Shapira D, Brot N, Weissbach H (1989) A member of the hsp70 family is localized in mitochondria and resembles *Escherichia coli* DnaK. Proc Natl Acad Sci USA 86: 7805–7808

Lewis MJ, Pelham HRB (1985) Involvement of ATP in the nuclear and nucleolar functions of the 70 kD heat shock protein. EMBO J 4: 3137–3143

Liberek K, Georgopoulos C, Zylicz M (1988) Role of the *Escherichia coli* DnaK and DnaJ heat shock proteins in the initiation of bacteriophage λ DNA replication. Proc Natl Acad Sci USA 85: 6632–6636

Lindquist S (1986) The heat-shock response. Ann Rev Biochem 55: 1151–1191

Lindquist S, Craig EA (1988) The heat-shock proteins. Ann Rev Genet 22: 631–677

Machamer CE, Rose JK (1988) Influence of new glycosylation sites on expression of the vesivular stomatitis virus G protein at the plasma membrane. J Biol Chem 263: 5948–5954

Marshall JS, DeRocher AE, Keegstra K, Vierling E (1990) Identification of heat shock protein hsp70 homologues in chloroplasts. Proc Natl Acad Sci USA 87: 374–378

McMullin TW, Hallberg RL (1987) A normal mitochondrial protein is selectively synthesized and accumulated during heat shock in *Tetrahymena thermophila*. Mol Cell Biol 7: 4414–4423

McMullin TW, Hallberg RL (1988) A highly evolutionary conserved mitochondrial protein is structurally related to the protein encoded by the *Escherichia coli groEL* gene. Mol Cell Biol 8: 371–380

Milarski KL, Morimoto RI (1989) Mutational analysis of the human hsp70 protein: distinct domains for nucleolar localization and adenosine triphosphate binding. J Cell Biol 109: 1947–1962

Milarski KL, Welch WJ, Morimoto RI (1989) Cell cycle-dependent association of hsp70 with specific cellular proteins. J Cell Biol 108: 413–423

Miziorko HM, Lorimer GH (1983) Ribulose-1,5-bisphosphate carboxylase-oxygenase. Ann Rev Biochem 52: 507–535

Mizzen LA, Chang C, Garrels JI, Welch WJ (1989) Identification, characterization, and purification of two mammalian stress proteins present in mitochondria, grp75, a member of the hsp70 family and hsp58, a homolog of the bacterial groEL protein. J Biol Chem 264: 20664–20675

Munro S, Pelham HRB (1986) An hsp70-like protein in the ER: identity with the 78 kd glucose-regulated protein and immunoglobulin heavy chain binding protein. Cell 46: 291–300

Munro S, Pelham HRB (1987) A C-terminal signal prevents secretion of luminal ER proteins. Cell 48: 899–907

Murakami H, Pain D, Blobel G (1988) 70-kD heat shock-related protein is one of at least two distinct cytosolic factors stimulating protein import into mitochondria. J Cell Biol 107: 2051–2057

Musgrove JE, Johnson RA, Ellis RJ (1987) Dissociation of the ribulosebisphosphate-carboxylase large-subunit binding protein into dissimilar subunits. Eur J Biochem 163: 529–534

Neidhardt FC, VanBogelen RA, Vaughn V (1984) The genetics and regulation of heat-shock proteins. Ann Rev Genet 18: 295–329

Nicchitta CV, Blobel G (1990) Assembly of translocation-competent proteoliposomes from detergent-solubilized rough microsomes. Cell 60: 259–269

Nicholson RC, Williams DB, Moran LA (1990) An essential member of the HSP70 gene family of *Saccharomyces cerevisiae* is homologous to immunoglobulin heavy chain binding protein. Proc Natl Acad Sci USA 87: 1159–1163

Normington K, Kohno K, Kozutsumi Y, Gething MJ, Sambrook J (1989) S. cerevisiae encodes an essential protein homologous in sequence and function to mammalian BiP. Cell 57: 1223–1236

Ostermann J, Horwich AL, Neupert W, Hartl FU (1989) Protein folding in mitochondria requires complex formation with hsp60 and ATP hydrolysis. Nature 341: 125–130

Patzer EJ, Schlossman DM, Rothman JE (1982) Release of clathrin from coated vesicles dependent upon a nucleoside triphosphate and a cytosol fraction. J Cell Biol 93: 230–236

Pedersen S, Bloch PL, Reeh S, Neidhardt FC (1978) Patterns of protein synthesis in *E. coli*: a catalogue of the amount of 140 individual proteins at different growth rates. Cell 14: 179–190

Pelham HRB (1984) HSP70 accelerates the recovery of nucleolar morphology after heat-shock. EMBO J 3: 3095–3100

Pelham HRB (1986) Speculations on the functions of the major heat shock and glucose-regulated proteins. Cell 46: 959–961

Pelham HRB (1988) Coming in from the cold. Nature 332: 776–777

Pelham HRB (1989a) Control of protein exit from the endoplasmic reticu um. Ann Rev Cell Biol 5: 1–23

Pelham HRB (1989b) Heat shock and the sorting of luminal ER proteins. EMBO J 8: 3171–3176

Phillips GJ, Silhavy TJ (1990) Heat-shock proteins DnaK and GroEL facilitate export of LacZ hybrid proteins in E. coli. Nature 344: 882–884

Picketts DJ, Mayanil CSK, Gupta RS (1989) Molecular cloning of a Chinese hamster mitochondrial protein related to the "chaperonin" family of bacterial and plant proteins. J Biol Chem 264: 12001–12008

Pinhasi-Kimhi O, Michalovitz D, Ben-Zeev A, Oren M (1986) Specific interaction between the p53 cellular tumour antigen and major heat shock proteins. Nature 320: 182–185

Pushkin AV, Tsuprun VL, Solovjeva NA, Shubin VV, Evstigneeva ZG, Kretovich WL (1982) High molecular weight pea leaf protein similar to the groE protein of Escherichia coli. Biochim biophys Acta 704: 379–384

Randall LL, Hardy SJS (1986) Correlation of competence for export with lack of tertiary structure of the mature species: a study in vivo of maltose-binding protein of E. coli. Cell 46: 921–928

Reading DS, Hallberg RL, Myers AM (1989) Characterization of the yeast HSP60 gene coding of a mitochondrial assembly factor. Nature 337: 655–659

Ritossa F (1962) A new puffing pattern induced by temperature shock and DNP in Drosophila. Experentia 18: 571–573

Rose JK, Doms RW (1988) Regulation of protein export from the endoplasmic reticulum. Ann Rev Cell Biol 4: 257–288

Rose MD, Misra LM, Vogen JP (1989) KAR2, a karyogamy gene, is the yeast homolog of the mammalian BiP/GRP78 gene. Cell 57: 1211–1221

Rothman JE (1989) Polypeptide chain binding proteins: catalysts of protein folding and related processes in cells. Cell 59: 591–601

Rothman JE, Kornberg RD (1986) An unfolding story of protein translocation. Nature 332: 209–210

Rothman JE, Schmid SL (1986) Enzymatic recycling of clathrin from coated vesicles. Cell 46: 5–9

Roy H, Cannon S (1988) Ribulose bisphosphate carboxylase assembly: What is the role of the large subunit binding protein? Trends Biochem Sci 13: 163–165

Ruben SM, VanDenBrink-Webb SE, Rein DC, Meyer RR (1988) Suppression of the Escherichia coli ssb-113 mutation by an allele of groEL. Proc Natl Acad Sci USA 85: 3767–3771

Sadis S, Raghavendra K, Schuster TM, Hightower LE (1990) Biochemical and biophysical comparison of bacterial dnaK and mammalian hsc73, two members of an ancient stress protein family. Curr Res Prot Chem (in press)

Sakakibara Y (1988) The dnaK gene Escherichia coli functions in initiation of chromosome replication. J Bacteriol 170: 972–979

Schleyer M, Neupert W (1985) Transport of proteins into mitochondria: translocational intermediates spanning contact sites between outer and inner membranes. Cell 43: 339–350

Schlossman DM, Schmid SL, Braell WA, Rothman JE (1984) An enzyme that removes clathrin coats: purification of an uncoating ATPase. J Cell Biol 99: 723–733

Smith SM, Ellis RJ (1979) Processing of small subunit precursor of ribulose bisphosphate carboxylase and its assembly into whole enzyme are stromal events. Nature 278: 662–664

Stevenson MA, Calderwood SK (1990) Members of the 70-kilodalton heat shock protein family contain a highly conserved calmodulin-binding domain. Mol Cell Biol 10: 1234–1238

Subjeck JR, Shyy T, Shen J, Johnson RJ (1983) Association between the mammalian 110,000-dalton heat-shock protein and nucleoli. J Cell Biol 97: 1389–1395

Sunshine M, Feiss M, Stuart J, Yochem J (1977) A new host gene (groPC) necessary for lambda DNA replication. Mol Gen Genet 151: 27–34

Tabita FR, McFadden BA (1974) D-Ribulose 1,5-diphosphate carboxylase from Rhodospirillum rubrum. J Biol Chem 249: 3459–3464

Tilly K, Georgopoulos C (1982) Evidence that the two Escherichia coli grcE morphogenetic gene products interact in vivo. J Bacteriol 149: 1082–1088

Ungewickell E (1985) The 70-kd mammalian heat shock proteins are structurally and functionally related to the uncoating protein that releases clathrin from coated vesicles. EMBO J 4: 3385–3391

VanDyk TK, Gatenby AA, Larossa RA (1989) Demonstration by genetic suppression of interaction of GroE products with many proteins. Nature 342: 451–453

Vogel JP, Misra LM, Rose MD (1990) Loss of BiP/GRP78 function blocks translocation of secretory proteins. J Cell Biol (in press)

Wada M, Itikawa H (1984) Participation of Escherichia coli K-12 groE gene products in the synthesis of cellular DNA and RNA. J Bacteriol 157: 694–696

Wada M, Sekine K, Itikawa H (1986) Participation of the dnaK and dnaJ gene products in phosphorylation of gultaminyl-tRNA synthetase and threonyl-tRNA synthetase of Escherichia coli. J Bacteriol 168: 213–220

Wada M, Fuhita H, Itikawa H (1987) Genetic suppression of a temperature-sensitive groES mutation by an altered subunit of RNA polymerase of Escherichia coli K-12. J Bacteriol 169: 1102–1106

Waegemann K, Paulsen H, Soll J (1990) Translocation of proteins into isolated chloroplasts requires cytosolic factors to obtain import competence. FEBS Lett 261: 89–92

Waldinger D, Eckerskorn C, Lottspeich F, Cleve H (1988) Amino acid sequence homology of a polymorphic cellular protein from human lymphocytes and the chaperonins from Escherichia coli (groEL) and chloroplasts (Rubisco-binding-protein). Biol Chem Hoppe-Seyler 369: 1185–1189

Waldinger D, Subramanian AR, Cleve H (1989) The polymorphic human chaperonin protein HuCha60 is a mitochondrial protein sensitive to heat shock and cell transformation. Eur J Cell Biol 50: 435–441

Walter G, Carbone A, Welch WJ (1987) Medium tumor antigen of polyma-virus transformation-defective mutant NG59 is associated with 73-kilodalton heat-shock protein. J Virol 61: 405–410

Wang C, Lazarides E (1984) Arsenite-induced changes in methylation of the 70,000 dalton heat-shock proteins in chicken embryo fibroblasts. Biochem Biophys Res Commun 119: 735–743

Welch WJ, Feramisco JR (1985) Rapid purification of mammalian 70,000-dalton stress proteins: affinity of the proteins for nucleotides. Mol Cell Biol 5: 1229–1237

Welch WJ, Suhan JP (1985) Morphological study of the mammalian stress response: characterization of changes occurring in cytoplasmic organelles, cytoskeleton and nucleoli, and appearance of intranuclear actin filaments in rat fibroblasts after heat-shock treatment. J Cell Biol 101: 1198–1211

Werner-Washburne M, Stone D, Craig EA (1987) Complex interactions among members of an essential subfamily of HSP70 genes in Saccharomyces cerevisiae. Mol Cell Biol 7: 2568–2577

Wiech H, Sagstetter M, Müller G, Zimmermann R (1987) The ATP requiring step in assembly of M13 procoat protein into microsomes is related to preservation of transport competence of the precursor protein. EMBO J 6: 1011–1016

Zimmermann R, Sagstetter M, Lewis MJ, Pelham HRB (1988) Seventy-kilodalton heat shock proteins and an additional component from reticulocyte lysate stimulate import of M13 procoat protein into microsomes. EMBO J 7: 2875–2880

Zylicz M, Georgopoulos C (1984) Purification and properties of the Escherichia coli dnaK replication protein. J Biol Chem 259: 8820–8825

Zylicz M, LeBowitz, McMacken R, Georgopoulos C (1983) The dnaK protein of Escherichia coli possesses an ATPase and autophosphorylating activity and is essential to an in vitro DNA replication system. Proc Natl Acad Sci USA 80: 6431–6435

Zylicz M, Ang D, Georgopoulos C (1987) The grpE protein of Escherichia coli. J Biol Chem 262: 17437–17442

Response of Mammalian Cells to Metabolic Stress; Changes in Cell Physiology and Structure/Function of Stress Proteins

W. J. WELCH, H. S. KANG, R. P. BECKMANN, and L. A. MIZZEN

1 Introduction

The ability of organisms to confront and survive adverse changes in their environmental circumstance represents an integral aspect of evolution. Throughout the animal kingdom there exist numerous examples of organisms which have evolved unique ways which allow for their continued survival in an environment that for the most part appears hostile and incompatible with the sustenance of life. In contrast to unique changes which occur in the whole animal to allow for acclimation to a particular environmental niche, cells from all organisms appear to employ a rather common and ubiquitous defensive mechanism when confronted with abrupt changes in their local environmental circumstance. This response, referred to as the heat shock or stress response, entails the rapid and coordinated increased expression of a group of proteins referred to as the heat shock or stress proteins. Collectively the stress proteins appear to provide the cell with protection during and/or after recovery from the environmental insult. That the stress response was developed at a very early step

Departments of Medicine and Physiology, University of California, Box 0854, San Francisco, CA 94143, USA

in evolution is supported by the fact that the stress proteins from prokaryotes through to higher eukaryotes all exhibit a considerably high degree of sequence homology (see other reviews by ASHBURNER and BONNER 1979; CRAIG 1985; LINDQUIST 1986).

A variety of studies provide support for the idea that the stress response and stress proteins are essential for the survival of the cell confronted with an environmental insult. First, by genetic means in bacteria and yeast, deletion of genes which encode the major stress proteins often renders the cells unable to survive a hyperthermic challenge (ITIKAWA and RYU 1979; CRAIG and JACOBSEN Similarly, in mammalian cells, injection of antibodies specific to stress proteins results in inability of the cells to survive a heat shock treatment (RIABOWOL et al. 1988). The second line of evidence indicating a defensive role for the stress response follows from studies examining a phenomenon referred to as "thermotolerance." Specifically if cells are subjected to a mild, sublethal heat shock treatment and then allowed to recover for a period of time at the normal growth temperature, they exhibit significantly higher survival rates when confronted with a second and what would otherwise be a lethal heat shock challenge (GERNER and SCHEIDER 1975; HENLE and LEEPER 1976). Thermotolerance appears transient and likely depends, at least in part, on the expression and function of the stress proteins (HENLE and LEEPER 1982; LI and WERB 1982; LANDRY et al. 1982; SUBJECK et al. 1982). However, other cellular changes elicited by the heat shock treatment (such as the arrest of the cell cycle) may also contribute to the thermotolerant state.

2 Induction of the Response

In addition to elevated physiological temperatures, mammalian cells initiate a stress response when exposed to a variety of other adverse conditions. Examples include (a) amino acid analogs, (b) various heavy metals, (c) high concentrations of ethanol, (d) various ionophores, (e) some inhibitors of mitochondrial function, and (f) thiol reactive agents, to name just a few (see NOVER 1984 for a list of inducers). In addition some (but not necessarily all) of the stress proteins are observed to increase in cells following (a) anoxia and/or release from anoxia, (b) viral infections, (c) changes in the levels of essential components such as glucose or calcium, and (d) agents which inhibit protein posttranslational modifications and/or protein trafficking (reviewed in SUBJECK and SHYY 1986).

The exact mechanism by which the cell recognizes a change in its environmental circumstances and activates the stress response is still unclear. One possible common denominator shared by the many treatments/agents which induce the response is their ability to promote protein denaturation and/or aggregation (reviewed in EDINGTON et al. 1989). Indeed, introduction of denatured or cross-linked proteins into frog oocytes is sufficient to trigger the stress response (ANANTHAN et al. 1986). Whether the accumulation in the cell of abnormally folded proteins is the only mechanism by which the stress response is initiated remains questionable. Moreover, whether all the agents/treatments which induce the response act via independent pathways or alternatively, converge within a single point along one pathway is unclear. Until more is known concerning the structure/function of the transcription factor(s) which regulate the expression of the stress genes, the answer to this interesting question of how the cell recognizes and activates the stress response will remain unsolved.

3 Changes in Cell Physiology After Stress

Not surprisingly, the cell under stress becomes refractory to other external stimuli. After heat shock an arrest in the cell cycle and a cessation of most activates involved with cellular proliferation are observed (RAO and ENGELBERG 1965; WESTRA and DEWEY 1971; ZEUTHEN 1971). In addition, changes in plasma membrane mediated events are observed including: (a) alterations in the activities of the Na^+/K^+ ATPase, thereby resulting in changes in the normal levels of intracellular Na^+/K^+ (YI 1979; BURDON and CUTMORE 1982; ANDERSON and HAHN 1985); (b) decreases in insulin and epidermal growth factor binding activities (CALDERWOOD and HAHN 1983; MAGUN and FENNIE 1981); (c) a reduction in the number of cell surface histocompatibility antigens (MEHDI et al. 1984); (d) a reduced ability of the lectin, concanavalin A, to induce plasma membrane protein patching and capping (STEVENSON et al. 1981); and (e) an increase in hexose transport activities into the cell (WARREN et al. 1986). Whether these changes are an indirect effect due to the inactivation (or denaturation) of proteins involved in these processes, or alternatively, represent a conscious effort of the stressed cell to downregulate or modify plasma membrane mediated events as part of a defensive mechanism is unclear.

An early consequence of hyperthermia is a rapid decrease in intracellular pH (WEITZEL et al. 1985; DRUMMOND et al. 1986), reduced levels of ATP (LEENDERS et al. 1974; ELLGAARD and MAXWELL 1975; FINDLY et al. 1983), and an increase in cytosolic calcium levels (STEVENSON et al. 1986; DRUMMOND et al. 1986). Because changes in pH and Ca^{2+} have been implicated as possible mediators by which certain "second messengers" are activated in the cell (e.g., those leading to changes in gene expression), DRUMMOND et al. (1986) examined whether such changes may be involved in the activation of the stress response. Blocking changes in pH after heat shock did not inhibit the increased synthesis of stress proteins. Similarly, raising intracellular calcium levels by other means did not turn on the high level expression of the stress proteins.

Owing to the fact that intracellular ATP levels drop after heat shock and because agents which interfere with mitochondrial function induce the response, it was suggested that changes in ATP production might be the trigger underlying activation of the response (discussed in ASHBURNER and BONNER 1979). Again, no direct evidence was found to support this idea. It is clear, however, that heat shock treatment results in a compromise in mitochondrial function and a subsequent increased dependence of the cell on anaerobic metabolism (MONDOVI et al. 1969; HAMMOND et al. 1982). Moreover, the shift from aerobic to anaerobic metabolism may be cruicial for the cell to survive a hyperthermic challenge. For example, yeast cells provided with only dextrose as a carbon source (which is utilized exclusively via glycolysis) appear more thermoresistant than do the cells growing on acetate (a metabolite used only via the respiration pathway) (LINDQUIST et al. 1982). Moreover, the levels of two glycolytic enzymes, enolase and glyceraldehyde 3-phosphate dehydrogenase, increase after heat shock treatment, most likely to facilitate the cells' increased dependence on glycolytic energy metabolism (IIDA and YAHARA 1985; NICKELLS and BROWDER 1988).

4 Morphological Changes After Stress

A number of morphological alterations have been described in mammalian cells following heat shock treatment. Within the nucleus increases in so-called perichromatin granules, which likely represent aggregates of unprocessed RNA, are observed shortly after temperature elevation (HEINE et al. 1971; WELCH and SUHAN 1985). These granules accumulate due to a failure of the cell to properly process (i.e., splice) mRNA and/or rRNA (MAYRAND and PEDERSON 1983; YOST and LINDQUIST 1986). In this regard it is important to note that at least for the major induced stress protein, hsp72, the corresponding gene contains no intervening sequences and therefore is not dependent upon splicing events for its synthesis and transport into the cytoplasm (see CRAIG 1985; LINDQUIST 1986 for review). Heat shock also inhibits nucleolar function, as evidenced by a block in ribosome biogenesis and a corresponding increase in denatured or aggregated preribosomal particles (SIMARD and BERNHARD 1967; AMALRIC et al. 1969; ELLGAARD and CLEVER 1971; RUBIN and HOGNESS 1975; NEUMANN et al. 1984; WELCH and SUHAN 1985). As discussed below, the nucleolus and, in particular, the aggregated preribosomes appear to be a major target for the most highly induced stress protein, hsp72 (WELCH and FERAMICSO 1984; PELHAM 1984; WELCH and SUHAN 1986). Finally, a third and rather curious abnormality is the appearance of actin containing filaments distributed throughout the nucleus after heat shock (PEKKALA et al. 1984; WELCH and SUHAN 1985; IIDA et al. 1986). These bundles of filamentous actin are observed to traverse the nucleus, but not to extend through the nuclear envelope. The function or relevance of these unusual intranuclear actin rods is entirely unclear.

A number of lesions in the cytoplasm are observed in the cell following hyperthermic treatment. Within minutes after heat shock the intermediate filaments are observed to relocalize or "collapse" into a meshwork of filaments closely enveloping the nucleus (FALKNER et al. 1981; THOMAS et al. 1982; WELCH et al. 1985; COLLIER and SCHLESINGER 1986). However, a similar collapse of the intermediate filaments occurs in response to other agents which do not result in the elevated expression of the stress proteins, and therefore, intermediate filament collapse is not likely to represent the "triggering" mechanism by which the response is initiated (WELCH and FERAMISCO 1985a). Upon return of the stressed cell back to its normal growth conditions the intermediate filaments slowly regain their normal well-spread distribution throughout the cell. Again, the relevance of these changes in intermediate filament morphology as it relates to the physiology of the stressed cell is unclear.

Accompanying the collapse of the intermediate filaments is the relocalization of both the mitochondria and polyribosomes into the perinuclear region of the stressed cell (WELCH and SUHAN 1986). In the electron microscope significant alterations in the integrity of various mitochondrial components are observed, including a swelling of the individual cristae and an enlargement of intra-cristal spaces (WELCH and SUHAN 1985). These changes in the integrity of the mitochondria appear similar to those observed after treatment of cells with various mitochondrial poisons which, like heat shock treatment, results in a reduction in ATP levels (WEINBACH and GARBUS 1968; BUFFA et al. 1970).

Finally, disruption of the Golgi complex is observed after heat shock. In contrast to its normal morphology of closely stacked membranes, heat shock treatment results in the fragmentation and dispersal of the Golgi complex into a collection of small vesicles distributed throughout the cytoplasm (WELCH and SUHAN 1985; ARRIGO et al. 1988). Whether such alterations affect the proper glycosylation and trafficking of secretory proteins has not been carefully examined.

In summary, the cell under stress redirects its outlook from one of maintenance or growth to one of defense. This is illustrated by the fact that after stress, the cell becomes refractory to most other external stimuli. The cell under stress experiences changes in a number of essential metabolic pathways, including a decreased ability to properly process RNA, a reduction in overall protein synthesis activities, an inability to manufacture new ribosomes, and compromises in overall mitochondrial function. Such pathways are perturbed presumably due to the inactivation (or denaturation) of key proteins/enzymes. Consequently, we suspect that one function of the stress proteins concerns the repair and recovery of such altered pathways.

5 The Mammalian Stress Proteins

In defining the stress proteins we often divide the family into two major groups: those referred to as the heat shock proteins (hsp's) and those referred to as the glucose regulated proteins (grp's). In lieu of a defined function we refer to the

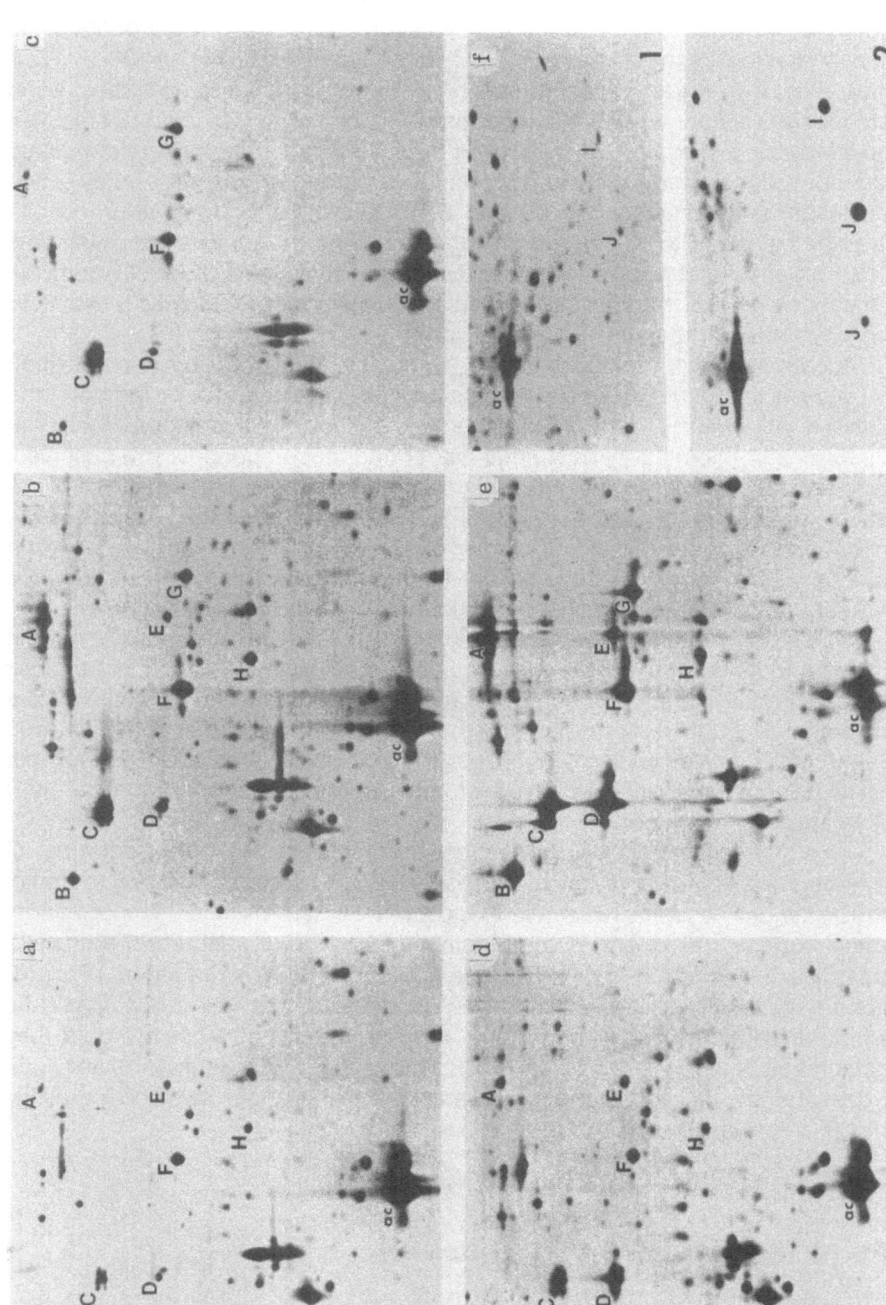

individual stress proteins according to their apparent molecular mass as determined by SDS-PAGE. Unfortunately, the stress proteins are often ascribed different molecular weights by different laboratories. Cells exposed to either elevated growth temperatures or heavy metals exhibit increased synthesis of proteins with apparent masses of approximately 8, 28, 32, 47, 58, 72, 73, 90, and 110 kilodaltons (kD) and a concomitant reduced expression of proteins whose synthesis was active prior to the insult (Fig. 1). These are referred to as the heat shock proteins. The glucose regulated proteins exhibit increased synthesis in cells (a) deprived of glucose, (b) treated with agents which perturb calcium homeostasis (EGTA or calcium ionophores), (c) treated with agents which inhibit proper glycosylation and/or prevent normal protein trafficking events, and (d) subjected to anoxia (reviewed in Subjeck and Shyy 1986). The grp's exhibit apparent masses of 75, 80, and 100 kD. In some cell types and depending upon the agent used (e.g., amino acid analogues) coordinated increased synthesis of both the hsp's and the grp's is observed (Fig. 1). Analysis of the stress proteins by two-dimensional gels demonstrates that: (a) most of the major stress proteins are relatively acidic polypeptides (pH 5–7), (b) most are synthesized at modest or even high levels in the normal unstressed cell, and (c) synthesis of other cellular proteins is reduced in the cell experiencing stress (Fig. 1).

Biochemical, immunological, and cDNA sequence analyses have demonstrated that the members of the two stress protein families are interrelated. For example, the grp75 and 80-kD proteins exhibit sequence homology with the 72- and 73-kD hsp's (Munro and Pelham 1986; Ting and Lee 1988; Craig et al. 1987, 1989). In addition, all of these proteins bind ATP (Welch and Feramisco 1985b; Mizzen et al. 1989). Similarly, grp100 is homologous to hsp90 (Sargan et al. 1986; Rebbe et al. 1987; Hickey et al. 1989). As will be discussed below, these related stress proteins appear to function in a similar fashion, but within their own distinct subcellular compartment.

5.1 Properties of the Individual Stress Proteins

Ubiquitin (8 kD). Levels of ubiquitin increase in both chickens and yeast after heat shock treatment, but so far it has not been reported as a heat-inducible protein in mammalian cells (Bond and Schlesinger 1986). Ubiquitin's most characterized function involves nonlysosomal protein degradation. Specifically,

---

Fig. 1 A–F. Stress proteins of mammalian cells. Rat embryo fibroblasts were labeled with [^{35}S]methionine: **A** at 37°C; **B** after a 43°C/90-min heat shock; **C** after exposure to 120 μM sodium arsenite; **D** after exposure to 10 μM calcium ionophore A223187; **E** after exposure to the proline analogue L-azetidine-2-carboxylic acid (5 mM). After labeling, the cells were harvested and the labeled proteins analyzed by two-dimensional gel electrophoresis (pH 4-3, acid end to the left). Shown in panels **A–E** are only those regions analyzing the higher molecular weight stress proteins. In panel **F** is a portion of the gels showing the lower molecular weight heat shock proteins (*1*, normal; *2*, arsenite treated cells). Stress proteins indicated are: *A*, 110 kD; *B*, 100 kD; *C*, 90 kD, *D*, 80 kD; *E*, 75 kD; *F*, 73 kD; *G*, 72 kD; *H*, 58 kD; *I*, 32 kD; *J*, 28 kD. Actin is indicated as *ac*

many proteins destined for turnover are first covalently modified by ubiquitin conjugation. Once modified, the conjugated protein is subsequently degraded by a nonlysosomal protease. Increased ubiquitin in the cell after stress presumably facilitates the removal of increasing amounts of denatured protein aggregates.

hsp28. In mammalian cells only a single low molecular heat shock protein of approximately 28 kD has been described (KIM et al. 1984; WELCH 1985). The protein is often overlooked in many studies owing to its relatively low content of methionine residues (HICKEY et al. 1986). In cells maintained at 37°C the steady state level of 28 kD is relatively low, with the protein being comprised of at least four major isoforms, three of which contain phosphate (KIM et al. 1984; WELCH 1985). Interestingly, within minutes after exposure of quiescent cells to various growth factors, cytokines, certain tumor promoters, or calcium ionophores, 28 kD phosphorylation (but not synthesis) is observed to increase as much as tenfold (WELCH 1985; KAUR et al. 1989; ARRIGO 1990). These changes in the phosphorylation state of 28 kD may be analogous to the situation in *Drosophila*, where the related multiple low molecular weight heat shock proteins exhibit differential expression and/or phosphorylation during development and differentiation (SIROTKIN and DAVIDSON 1982; CHENEY and SHEARN 1983; ZIMMERMAN et al. 1983). Like in *Drosophila*, mammalian hsp28 levels increase in some cells exposed to steroids (IRELAND and BURGER 1982; EDWARDS et al. 1980). A cDNA complementary to 28 KD mRNA has been cloned and sequenced (HICKEY et al. 1986; MORETTI-ROJAS et al. 1988). Like the low molecular weight hsp's from other organisms, mammalian hsp28 exhibits homology to the lens α-crystallin proteins.

hsp28 purified from HeLa cells exists as a large homo-oligomer with a native mass of approximately 400 kD, and in the electron microscope the purified protein exhibits a "doughnut-like" appearance (ARRIGO and WELCH 1987). After heat shock the protein further aggregates into complexes as large as 2×10^6 daltons (ARRIGO et al. 1988). Because the related α-crystallin proteins form similar large aggregates, we suspect that those domains shared in common between the two proteins may be responsible for their similar self-assembling properties. Indirect immunofluorescence data have revealed a perinuclear distribution of 28 kD, in close proximity to the Golgi complex in the 37 °C cell. After heat shock treatment, however, much of hsp28 moves into the nucleus, where it is found as very large aggregates (ARRIGO et al. 1988). During recovery of the cells back at 37 °C the hsp28 nuclear complexes begin to disaggregate and the protein eventually returns to the cytoplasm.

The biochemical function of hsp28 remains unclear. Our own hypothesis is that hsp28 may be important in some aspect of cell growth control. We suggest this since hsp28 levels increase significantly in cells experiencing metabolic stress, a condition which results in cell cycle arrest. Conversely, in cells stimulated to proliferate by addition of various mitogens, hsp28 phosphorylation (but not synthesis) increases. Two other observations reinforce the idea that hsp28 may be important in some aspect of growth control. First, in *Drosophila*, during early embryogenesis synthesis of the analogous low molecular weight stress proteins is not observed. During later stages of development, however, they are constitutively expressed (SIROTKIN and DAVIDSON 1982; CHENEY and SHEARN 1983; ZIMMERMAN et al. 1983). Interestingly, exposure of the early embryos to agents which inhibit subsequent development (i.e., teratogens) results in the inappropriate expression of these small heat shock proteins (BUZIN et al. 1982; BUZIN and BOURNIAS-VARDIABASISAIN 1982). Secondly, in mammalian cells an inverse relationship between hsp28 expression and cellular oncogenicity has been observed. For example, in a variety of adenovirus transformed rat cells, hsp28 expression appears absent in only those cell lines which exhibit tumorigenic potential. Moreover, in such cell lines, hsp28 expression is not increased after heat shock treatment (ZANTEMA et al. 1989). These properties of the low molecular weight heat shock proteins appear reminiscent of those exhibited by so-called antioncogene proteins. Specifically, inappropriate expression of the *Drosophila* low molecular weight hsp's during early embryogenesis results in an arrest of development. In mammalian cells, a lack of hsp28 expression correlates with cellular oncogenic potential. Finally, phosphorylation of hsp28 increases in the response of cells to various mitogens. Consequently, we are examining whether hsp28 serves some role in the proliferative response, and whether phosphorylation of the protein is important in regulating its activity.

hsp32. Induction of hsp32 is most obvious in cells exposed to various heavy metals such as arsenite and cadmium or agents which react with free sulfhydryl groups (e.g., iodoacetimide), or in cells exposed to either ultraviolet light or the oxidizing agent, hydrogen peroxide (LEVINSON et al. 1980; CALTABIANO et al. 1986; KEYSE and TYRELL 1987). The gene encoding hsp32 has been isolated and shown to be equivalent to heme oxygenase (KAGEYMA et al. 1988). This protein is essential for the breakdown of heme, catalyzing the first step in the conversion of heme to its final product, bilirubin. The question arises then as to why the cell would require increased levels of heme oxygenase in response to various oxidative stressors. A clue to the answer lies in the fact that heme is an essential component of a number of important enzymes such as the cytochromes. Considering the fact that overall mitochondrial function, specifically ATP production, is compromised in the cell exposed to various stress agents, one wonders whether perturbations in cytochrome structure/function may be a contributing factor. If so, increases in heme oxygenase might be necessary to facilitate the turnover of heme-containing proteins, like the cytochromes, which may have become denatured and inactivated as a consequence of the stress event.

47 kD. NAGATA and colleagues have described a heat induced 47-kD protein in chicken fibroblasts (NAGATA et al. 1986). This protein has an unusually basic pI of approximately 9.0, and exhibits an affinity for collagen in a pH dependent manner. Curiously, this collagen binding protein appears to be present near or within the endoplasmic reticulum. Following transformation of chick fibroblasts with Rous sarcoma virus, synthesis of 47 kD decreases approximately two- to threefold, yet its level of phosphorylation appears to increase five- to sevenfold (NAGATA and YAMADA 1986). In contrast, although the synthesis of 47 kD increases after heat shock treatment, the relative levels of 47 kD phosphorylation do not change. Further studies are needed to ascertain whether a similar protein exists in other eukaryotic cells and to define the function of this somewhat novel stress protein.

hsp58 (or GroEL homologue). hsp58 represents the mammalian counterpart of the bacterial GroEL heat shock protein. Genetic and biochemical studies have shown that *E. coli* GroEL is required for the morphogenesis of bacteriophage λ. Here the protein facilitates the correct posttranslational assembly of the bacteriophage head and tail protein complexes (STERNBERG 1973; COPPO et al. 1973; GEORGOPOULOS et al. 1973; GEORGOPOULOS and HOHN 1978). A protein present within plant chloroplasts has been isolated and shown to be similar in sequence to GroEL (HEMMINGSEN et al. 1988). This protein, termed the Rubisco binding protein, facilitates the proper posttranslational assembly of Rubisco, an oligomeric enzyme involved in CO_2 fixation in all higher plants (BOCHKANEVA et al. 1988; GOLOUBINOFF et al. 1988). Finally, a GroEL-like protein has also been reported in both *Tetrahymena* and yeast (MCMULLEN and HALLBERG 1987; CHENG et al. 1989). In both cases the protein, with an apparent mass of approximately 58–60 kD, is synthesized in the cytosol as a precursor and is transported into the mitochondria. Within the mitochondrial matrix, hsp58 is thought to function in facilitating the proper assembly of various multimeric mitochondrial enzyme complexes. Thus the bacterial GroEL, the plant Rubisco binding protein, and the *Tetrahymena* and yeast hsp60 all appear to represent a related class of proteins which function to facilitate or "chaperone" the assembly of monomeric proteins into higher ordered oligomeric structures. Consequently, this family of related proteins has been referred to as "molecular chaperones" or "chaperonins" (HEMMINGSEN et al. 1988) and are described in more detail in this volume by LANGER and NEUPERT.

We have recently purified and partially characterized the mammalian counterpart to GroEL, hsp58 (MIZZEN et al. 1989). Like in *Tetrahymena* and yeast, mammalian hsp58 is synthesized in precursor form and proteolytically processed during or after its import into the mitochondria. Gel filtration or sucrose velocity sedimentation analyses revealed the protein to be an oligomer with a native molecular mass of approximately 500–600 kD (MIZZEN and WELCH, unpublished observations). Antibodies specific to hsp58 stain mitochondria, and preliminary data indicate the protein to be situated within the mitochondrial matrix (MIZZEN et al. 1989). Owing to its relationship to the so-called chaperonins we suspect that mammalian hsp58 functions to facilitate proper oligomeric assembly of proteins within the matrix of the mitochondria.

110 kD. Of all the major members of the stress protein family, we know the least regarding the properties of this protein. 110 kD is a constitutively expressed protein whose synthesis increases approximately fivefold after stress. The protein appears to be present within that region of the nucleolus involved in rRNA transcription and/or processing (SUBJECK et al. 1983). Considering that rRNA transcription is inhibited after heat shock, one wonders whether 110 kD is involved in the resumption of normal nucleolar transcription events during recovery from the heat shock insult.

The hsp90 Family. As was discussed earlier, two of the stress proteins, hsp90 and grp100, share sequence homology and therefore we suspect the two proteins will have similar biochemical functions, but within their own distinct intracellular compartments. Specifically, hsp90 is primarily a cytosolic protein while grp100 is compartmentalized within the endoplasmic reticulum, Golgi complex, and perhaps on the plasma membrane. The properties of these two proteins are outlined below.

hsp90. This protein is also referred to as hsp83 or hsp89 and is the mammalian equivalent of the *Drosophila* 83-kD stress protein. Mammalian hsp90 is a very abundant protein in all cells grown under normal conditions and its synthesis increases three- to fivefold after heat shock treatment (WELCH et al. 1983). In cells deprived of glucose or oxygen, or treated with agents which perturb calcium homeostasis (e.g., EGTA or calcium ionophores), synthesis of hsp90 declines concomitant with an increased synthesis of the major grp's, 75, 80, and 100 kD (WELCH et al. 1983). hsp90 is an extremely heterogeneous protein with there being at least 12 isoforms, at least half of which are phosphorylated (WELCH et al. 1983). In mammalian cells there appear to be two genes which encode hsp90, α and β, which appear approximately 70% related (REBBE et al. 1987; HICKEY et al. 1989). When purified to homogeneity hsp90 exists as a dimer with rod-like geometry (WELCH and FERAMISCO 1982). It is not yet clear whether the protein exists as a hetero- or homodimer of the α and β subunits.

Although the purified protein exists as a dimer, analysis of whole cell lysates by gel filtration reveals an extremely heterogeneous profile of hsp90, perhaps indicative of its interaction with other cellular proteins (WELCH, unpublished observations). Indeed, previous studies have demonstrated an interaction of hsp90 with a number of rather interesting macromolecules. For example, hsp90 along with another cellular protein, p50, forms transient interactions with a number of oncogenic tyrosine kinases, the most well studied being pp60[src] (see BRUGGE 1989 for review). In cells transformed with Rous sarcoma virus, newly synthesized pp60[src], present in the cytoplasm, rapidly associates with hsp90 and p50, and while in this complex pp60[src] exhibits no tyrosine kinase activity or autophosphorylated tyrosine residues. As the complex containing pp60[src] moves to the plasma membrane, the complex dissociates; pp60[src] is deposited at the inner side of the plasma membrane and now exhibits both its tyrosine kinase activity and autophosphorylated tyrosine residues. Hence hsp90 (and perhaps p50) appears to bind to newly synthesized pp60[src] to maintain the kinase in an inactive form until its translocation to the plasma membrane. In a seemingly analogous scenario, hsp90 appears to regulate the activity of various steroid hormone receptors (CATELLI et al. 1985; SANCHEZ et al. 1985). Almost all steroid hormone receptors (estrogen, testosterone, progesterone, gluococorticoid) exist within the cytosol or nucleus in complex with hsp90. This form of the receptor, termed the 9S or "nontransformed" receptor, is unable to bind DNA (reviewed by BAULIEU 1988). Addition of steroid, which freely passes through the plasma membrane, leads to binding to the steroid receptor protein and the dissociation of hsp90 from the receptor. The receptor, now in its activated or "transformed" 4S form, can bind to its appropriate target gene and elicit transcription events. Thus, again we have a situation in which hsp90 appears to maintain a target protein, in this case the steroid receptor, in an inactive form, presumably by preventing DNA binding in the absence of steroid hormone. Finally, hsp90 copurifies with the so-called heme regulated protein kinase, whose substrate is the alpha subunit of eukaryotic initiation factor 2 (eif-2α) (ROSE et al. 1989). eif-2α is important in regulating translational initiation events and when phosphorylated prevents the initiation of protein synthesis (DUNCAN and HERSHEY 1984). While hsp90 does interact with the eif-2α kinase, its exact role in regulating kinase activity is unclear. Nevertheless, the observation that eif-2α exhibits increased phosphorylation after heat shock may explain why overall protein synthesis activities are transiently arrested in such cells. In summary, all of the available evidence points toward a role for hsp90 in regulating the activity of target proteins through which it transiently interacts. Somewhat confusing, however, is the fact that the levels of hsp90 in the cell are significantly higher than would be required in regulating the activities of only tyrosine kinases, steroid receptors, and the eif-2α kinase, proteins which are present in relatively low levels. Hence it will be interesting to see whether this unusual stress protein participates in regulating the activities of other proteins/enzymes and whether such regulation is affected as a consequence of heat shock or other stress treatments.

grp100 (also referred to as grp94). The related homologue of hsp90, grp100, is a compartmentalized protein present within the endoplasmic reticulum (ER) LEWIS et al. 1985; KOCH et al. 1986). A number of studies also have reported grp100 within the Golgi complex and perhaps on the plasma membrane (POUYSSEGUR et al. 1977; WELCH et al. 1983; OLDEN et al. 1978; MCCORMICK et al. 1979). The protein is posttranslationally modified by the addition of both phosphate and carbohydrate moieties (POUYSSEGUR et al. 1977; WELCH et al. 1983). The predicted amino acid sequence of grp100 contains a putative transmembrane spanning domain within the first third of the molecule and therefore it has been suggested that the protein spans the ER/Golgi membrane with a considerable portion of the protein present on the cytosolic side of the membrane (MAZZARELLA and GREEN 1987). At the present time we know relatively little concerning the function of grp100. Within isolated microsomes grp100 appears sensitive to added protease, a result consistent with it being a transmembrane protein (KANG and WELCH, in preparation). grp100 binds calcium and such binding appears to affect the apparent size of the protein as assayed by gel filtration (KOCH et al. 1986; KANG and WELCH, in preparation). Owing to its homology with hsp90, we suspect that grp100 will exhibit transient interactions with other macromolecules. For example, studies are in progress to determine whether grp100 binds to certain proteins as they pass through the ER/Golgi complex.

5.2 The hsp70 Proteins: Related ATP Binding Proteins Distributed Within Different Intracellular Compartments

As was discussed earlier, there exist at least four major stress proteins which exhibit considerable sequence homology, bind ATP, and appear to function in a similar biochemical fashion, but within different subcellular compartments. This family of related stress proteins consists of the hsp73, hsp72 grp75, and grp80 proteins (Table 1). The properties of these proteins are outl ned below:

hsp72 and 73 (also referred to as hsp70 and 72). The two major hsp70 proteins are referred to in our laboratory as hsp72 and 73. Each prote n exhibits multiple isoelectric forms, the basis of which is unknown. hsp73 is present in relatively high levels in the normal, unstressed cell and therefore is often referred to as the "cognate" or "constitutive" hsp70 protein. hsp72 is highly related to 73 kD and is synthesized at very high levels after stress (reviewed in CRAIG 1985; LINDQUIST 1986). In most cells, synthesis of hsp72 occurs only after induction of the stress response. An exception to this rule occurs in primate cells. So far in all primate cells examined (over 15 human and three monkey cell lines), synthesis of hsp72 is observed in the cells grown under normal conditions (WELCH et al. 1983; and unpublished observation). A number of interesting observations concerning this constitutive expression of hsp72 in primate cells include the following: (a) its expression appears cell cycle regulated, with its synthesis observed at the G_1/S boundary (MILARSKI and MORIMOTO 1986); (b) 72 kD expression, in general, appears to increase in primate cells following transformation (WELCH, unpublished observation); and (c) the expression of 72 kD increases in human cells

Table 1. The hsp70 family

Member	Expression		Locale	Function
	Normal	Stress		
hsp73	+ +	+ + +	Cytoplasm/nucleus	Facilitates: protein folding/assembly; uncoating of clathrin vesicles; protein translocation; recognition, binding and turnover(?) of denatured proteins; DNA replication (?)
hsp72	±	+ + + + +	Cytoplasm/nucleus	Similar to hsp73. Other as yet to be defined functions (?)
grp78 (BiP)	+ +	+ + +	Endoplasmic reticulum	Facilitates: import, glycosylation, folding and assembly of proteins; recognition and removal of denatured/unfolded proteins; nuclear fusion in yeast
grp75	+	+ +	Mitochondria	Facilitates: import, folding, and assembly of proteins in matrix; DNA replication events
70 kD	+	+	Chloroplasts	Similar to grp75 in mitochondria
70 kD	+	+	Plasma membrane (?)	Facilitates: antigen presentation; conformational changes in plasma membrane proteins

(but not rodent cells) in response to transfection with either of two so-called cooperating viral oncogenes, *E1A* or *myc* (NEVINS 1982; KINGSTON et al. 1984; WU et al. 1986).

Immunological, biochemical, and DNA sequence analyses have demonstrated that while the 72-kD and 73-kD proteins are highly related, they are in fact distinct gene products (reviewed in CRAIG 1985). Both proteins exhibit very similar biochemical properties, including their stoichiometric copurification during either gel filtration or ion exchange chromatography, and both exhibit similar intracellular locales (WELCH and FERAMISCO 1982, 1985b; WELCH and MIZZEN 1988).

A number of observations have implicated a role for the 72/73-kD stress proteins in biochemical pathways essential to the lifestyle of normal, unstressed cells. First, in vitro, the 72/73-kD stress proteins facilitate the uncoating and release of clathrin triskelions from clathrin coated vesicles (UNGEWICKEL 1985; CHAPPELL et al. 1986). The 70-kD stress proteins bind to the clathrin coat and facilitate removal of the clathrin triskelion subunits. This uncoating activity appears to require ATP hydrolysis in order to displace each leg of the clathrin triskelion from the coated vesicle. These uncoating events do not appear to be truly catalytic since the 70-kD proteins remain bound to their substrate, the released triskelion (SCHLOSSMAN et al. 1984; SCHMID et al. 1985). hsp72 and 73 also appear important in the mechanism by which proteins are translocated across intracellular membranes in yeast. Here hsp72 and 73 appear to bind to and stabilize preproteins prior to their import into the mitochondria or the ER. Again the process requires ATP (and likely, ATP hydrolysis) and presumably one or more additional factors in the cell (CHIRICO et al. 1988; DESHAIES et al. 1988). It is not yet clear whether the 70-kD stress proteins actually participate directly in the translocation event or alternatively, maintain the target protein in a "translocation competent state" prior to its transfer across the membrane.

grp80 (also referred to as grp78 or BiP). grp80 is a relatively abundant protein in cells grown under normal conditions. The protein contains phosphate and/or is ADP-ribosylated and is present within the ER (CARLSSON and LAZARIDES 1983; ZALA et al. 1980 MUNRO and PELHAM 1986). grp80 exhibits approximately 50% sequence homology with the hsp70 proteins and similarly binds ATP (WELCH and FERAMISCO 1985b; MUNRO and PELHAM 1986; TING and LEE 1988). By sequence analysis grp80 has been shown to be equivalent to a previously described protein, BiP (MUNRO and PELHAM 1986). BiP, present within the lumen of the ER, interacts transiently with maturing IgG heavy (H) and light (L) chains, presumably to facilitate their assembly into the mature H_2L_2 molecule (HAAS and WABL 1983; BOLE et al. 1986; HENDERSHOT et al. 1987). Similarly, BiP binds to newly synthesized influenza virus hemagglutinin monomers and catalyzes their trimerization within the ER (GETHING et al. 1986). Thus, in general, grp80 (or BiP) appears to function, in an ATP dependent manner, to facilitate the stabilization, correct folding, and/or oligomeric assembly of proteins which pass through the ER. In addition, BiP may also serve a role in the recognition and/or removal of some proteins which are secreted into the ER but which fail to move further along the secretory pathway (KASSENBROCK et al. 1988; KOZUTSUMI et al. 1988).

grp75. Synthesis of grp75 is affected by many of the same treatments (e.g., glucose starvation, calcium ionophores) which increase the synthesis of the two other major glucose regulated proteins, grp80 and 100 (WELCH et al. 1983; MIZZEN et al. 1989). grp75 represents yet another member of the 70-kD family of stress proteins as evidenced by immunological cross-reactivity and its ability to bind ATP (MIZZEN et al. 1989). The protein is synthesized initially as precursor in the cytoplasm and is translocated into the mitochondria. On the basis of these observations we suspect that grp75 is the mammalian equivalent of the SSCI gene product of yeast (CRAIG et al. 1989). We suspect that grp75 within the mitochondria, like the grp80 protein present in the ER, serves a general role in the proper assembly and/or disassembly of target proteins. Recent evidence indicates that grp75, like its related counterpart *E. coli* DnaK, also is essential for mitochondrial DNA replication (T.W. WONG, personal communication).

Other hsp70 Homologues. It should be mentioned that hsp70 homologues have now been reported in plant chloroplasts and perhaps on the plasma membrane. For example, three related hsp70 proteins have been reported within plant chloroplasts. All are constitutively expressed and do not appear to increase after heat shock. Using biochemical fractionation one of these hsp70 proteins was shown to be associated with the outer envelope membrane while the other two homologues were found as soluble proteins present in the chloroplast stroma (MARSHALL et al. 1990). Finally, PIERCE and colleagues have reported that a form of hsp70 may be present on plasma membrane of some antigen presenting cells and may play a role in the process of antigen presentation (VANBUSKIRK et al. 1989).

5.2.1 The Cytosolic hsp70 Proteins Interact with Newly Synthesized Proteins

All of the available data indicate a role for the family of hsp70 proteins in facilitating protein maturation events. To reiterate, transient interactions of hsp70 proteins have been observed with various other proteins which are in the process of posttranslational translocation and/or assembly. However, considering the high constitutive expression of cytosolic hsp73 and the fact that agents/treatments which activate the stress response are protein denaturants, we suspected that hsp73 (and perhaps hsp72) may be serving a more general role in protein folding/assembly events. Consequently, we undertook studies examining for possible interactions of hsp72/73 with other cellular proteins. Using metabolic pulse-labeling and immunoprecipitation with hsp72/73 antibodies, we observed that most, if not all, newly synthesized proteins transiently interact with hsp72/73. Specifically, if cells are labeled with [^{35}S]methionine for only a 15-min period, most of the newly synthesized and radiolabeled proteins are found in a complex with hsp72/73. If the pulse-labeled cells are allowed a subsequent 2-h chase period in the absence of radiolabel, most of the radiolabeled proteins are no longer found complexed with hsp72/73. This interaction with newly synthesized proteins appears to be a cotranslational event since hsp72 and 73 appear to bind to the nascent chains of the maturing polypeptides. Once translation of the protein is complete, release of hsp72/73 occurs, presumably via the binding to and hydrolysis of ATP (BECKMANN et al. 1990). Similar experiments using cells treated with an amino acid analogue provided a similar, although distinct result. In cells exposed to the amino acid analogue of proline, Azc, again all of the newly synthesized proteins were found to coprecipitate with hsp72/73 antibodies. Unlike the situation with the normal cells, however, the interaction of hsp72/73 with the newly synthesized and analogue containing proteins was not transient. Even after 2h following their

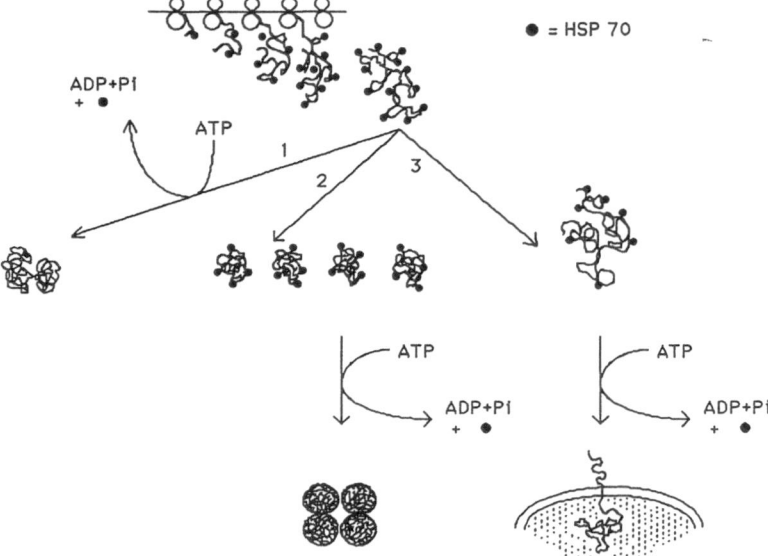

Fig. 2. A model by which hsp70 interacts with and facilitates maturation of newly synthesized proteins. During protein synthesis peptide domains emerge from the ribosome and interact with hsp70. Multiple copies of hsp70 may be necessary to interact with the nascent polypeptide domains, the number depending upon the size of the protein. Binding to hsp70 maintains the completed polypeptide in a semifolded, or perhaps entirely unfolded state. For monomeric proteins (1) folding occurs through the release of hsp70 proteins from the peptide domains as they fold together into their final conformation. Release of hsp72/73 is facilitated by ATP hydrolysis. For proteins which are to be assembled into an oligomeric structure (2), partial folding of the monomers with a concomitant release of some hsp70 molecules takes place. These "assembly competent" monomers now proceed to assemble into an oligomeric structure, accompanied by the release of remaining hsp70 proteins. For those proteins which are to be translocated from the cytosol into organelles (the ER or mitochondria) (3), the target protein is maintained in a semifolded, or unfolded, "translocation competent state" by virtue of the bound hsp70 protein. As translocation proceeds, the hsp70 protein(s) are released. Once translocated subsequent folding of the protein occurs, perhaps facilitated by other chaperonin-like molecules with the organelle. In cells placed under stress by exposure to amino acid analogue, hsp72/73 again binds to the nascent chains. In this case, however, the newly synthesized and analogue containing protein cannot properly fold, and hsp72/73 release does not occur. Consequently the free and available pool of hsp72/73 is reduced and the cell responds by increasing the synthesis of new hsp72/73

synthesis, the Azc containing proteins were still found in a stable complex with hsp72/73.

We interpret these studies to indicate a role for hsp72/73 in facilitating protein folding events (Fig. 2). As proteins are being translated, peptide domains emerge from the ribosome and bind to hsp72/73. Such an interaction may prevent premature folding of the nascent chain until its translation has been completed. Once completed, and now with perhaps multiple peptide domains bound to hsp72/73, folding of the polypeptide commences. Presumably the affinity of neighboring peptide domains as they fold together is greater than that for their binding to hsp72/73. Consequently, hsp72/73 is released, perhaps via hydrolysis of ATP. In the case of cells treated with amino acid analogues, again the newly synthesized proteins are found in a complex with hsp72/73. However, subsequent release of hsp72/73 from the newly synthesized and analogue containing proteins does not occur. Under these conditions we suspect that the target protein, which contains the analogue, cannot properly fold and consequently hsp72/73 release does not occur. With time these "abnormal" proteins are degraded by the cell. Perhaps the abnormally folded proteins remain bound to hsp72/73 to maintain their solubility until presentation to the appropriate proteolytic system. In addition to these transient interactions, we have also observed that some newly synthesized proteins remain stably bound to hsp72/73. Such stably bound proteins may represent a class of polypeptides which are to be translocated into organelles or which are to be assembled into higher ordered, multimeric structures (Fig. 2).

5.2.2 Regulation of hsp70 Synthesis

These results also may reflect on the mechanism by which the cell recognizes a stress event and subsequently regulates the increased expression of hsp72/73. For example, hsp72/73 likely exists in the cell in equilibrium between its free and substrate bound forms, the latter representing proteins in the process of folding or assembly. When the equilibrium is shifted toward the longer lived, substrate bound forms (as occurs in the Azc treated cells), the corresponding reduction in the amount of free hsp72/73 results in the activation of new hsp72/73 synthesis. Similarly, a reduction in the free pool of hsp72/73 may occur as a result of binding to proteins which have become denatured in the cell after heat shock, again thereby triggering synthesis of hsp72/73. Once the "free" pool of hsp72/73 is restored to some critical level, the response is turned off and the cells return to their normal pattern of protein synthesis. This mode of hsp72/73 regulation would also help explain a number of previous observations showing that: (a) hsp72 synthesis appears to be autoregulated, with a considerable part of the regulation occurring posttranslationally (DIDOMENICO et al. 1982); (b) hsp72 synthesis increases in direct proportion to the severity of the stress (DIDOMENICO et al. 1982; MIZZEN and WELCH 1988); (c) the preexisting levels of hsp72/73 influence the amount of new hsp72/73 synthesis following a second stress event (MIZZEN and WELCH 1983); (d) microinjection of denatured proteins into frog oocytes is sufficient to activate synthesis of hsp72, presumably due to binding of hsp72/73 to the denatured proteins (ANANTHAN et al. 1986); and (e) microinjection into cells of antibodies specific to hsp72/73 results in an immediate induction of new hsp72/73 synthesis (WELCH, unpublished observations). This mode of regulation likely exists for the hsp70 related protein, BiP. As was mentioned earlier, in situations where BiP exhibits longer lived interactions with target proteins unable to fold correctly, the cells respond by increasing new BiP synthesis.

5.2.3 Role of hsp70 Proteins in Cells Experiencing Metabolic Stress

Having discussed some of the properties and possible functions of the stress proteins in the normal cell, the obvious question arises as to their role in the cell experiencing stress. While we still do not have definitive answers to this question, progress is being made in this direction. Because hsp72 is expressed at low or negligible levels in the unstressed cell, but at very high levels after stress, most attention with regard to the biochemistry of the stressed cell has focused on hsp72.

After heat shock much of hsp72 is found within the nucleus and in particular the nucleolus (VELAZQUEZ and LINDQUIST 1984; WELCH and FERAMISCO 1984; PELHAM 1984). Only those nucleoli which appear severely altered contain hsp72, the protein presumably in complex with the denatured or aggregated peribosomes (WELCH and SUHAN 1985). Our current hypothesis is that hsp72 functions to facilitate repair of the damaged nucleolus, perhaps by either helping refold some of the unfolded ribosomal proteins or by solubilizing the denatured ribosomal proteins to facilitate their turnover. In light of our data regarding a role for hsp72/73 in facilitating protein folding and assembly events, we are examining the possibility that under normal cellular conditions, the hsp70 proteins may also be involved in the normal transport of ribosomal precursor proteins into the nucleolus and their subsequent assembly into the mature ribosomal particle. Heat shock treatment may block such assembly by causing denaturation of the ribosomal precursors. As a consequence hsp72/73 accumulates in the nucleolus in association with the denatured ribosome proteins.

During release from heat shock and as the nucleoli begin to recover their normal activities, much of hsp72 exits back into the cytoplasm. Here much of hsp72 and hsp73 is found colocalized with the translation machinery (VELAZQUEZ and LINDQUIST 1984; WELCH and SUHAN 1986). We suspect that such a colocalization may be due to the high level of protein synthesis occurring and the need for hsp72/73 in facilitating protein folding events. Within the cytoplasm hsp72/73 may also function in binding to proteins which have become denatured as a consequence of the thermal treatment (BECKMANN and WELCH, in preparation). In such situations hsp72/73 may act to stabilize the denatured (or unfolded) proteins and thereby allow for their subsequent refolding or, in the case of those proteins irreversibly denatured, facilitate their solubilization and eventual turnover.

In the case of the other stress proteins, their exact role in the cell experiencing stress is still somewhat unclear. As was mentioned earlier, we suspect that hsp28 may serve some function in regulating the proliferation capacity of the cell, perhaps being integral in some aspect of the cell cycle. Considering that heat shock treatment results in an arrest of the cell cycle and that such arrest is important for the survivability of the stressed cells, we are currently examining the possible role of hsp28 in mediating the arrest of the cell cycle. For hsp90, all of the available evidence implicates it being involved in the regulation of other important polypeptides (e.g., steroid hormone receptors, tyrosine kinases, eif-2α kinase). Hence, hsp90 may be important in redirecting the activity of its target proteins after heat shock. Finally, the hsp58 or GroEL homologue appears important in the assembly of multimeric proteins within the mitochondria. Because mitochondrial function is perturbed after heat shock one wonders whether this is due to the denaturation of critical mitochondrial enzyme complexes. If so, increased levels of hsp58 may facilitate either the reassembly of such denatured complexes or alternatively may be necessary to mediate the assembly of new proteins imported into the mitochondria to replace those denatured by the stress event.

6 Summary

In response to adverse changes in their local environment, cells or tissues from all organisms increase the expression of a group of proteins referred to as heat shock or stress proteins. Collectively, the stress proteins are thought to provide the cell with some degree of protection during the environmental insult as well as facilitate the repair and recovery of metabolic pathways perturbed as a consequence of the stress event. Within the past few years it has become apparent that most all of the stress proteins are present in appreciable levels in the unstressed cell and are involved in a number of very basic and essential biochemical pathways. The present review has discussed pertinent changes in cell physiology in mammalian cells experiencing metabolic stress. In addition, considerable attention has been given to discussing the properties and possible functions of the individual stress proteins.

Acknowledgments. This work was supported by N.I.H. grant GM33551 (W..W.), an MRC postdoctoral fellowship (L.A.M.), a KOSEF fellowship (H.S.K.), and Department of Medicine funds (R.P.B.). The authors thank M. Lovett and A. Kabiling for excellent technical assistance and T. Kleven for preparation of the manuscript.

References

Amalric F, Simard R, Zalta JP (1969) Effect de la temperature supraoptimale sur les ribonucleo-protines. II Etude Biochimique Exp Cell Res 55: 370–377

Ananthan J, Goldberg AL, Voellmy R (1986) Abnormal proteins serve as eukaryotic stress signals and trigger the activation of heat shock genes. Science 232: 252–254

Anderson R, Hahn GM (1985) Differential effects of hyperthermia on the Na$^+$; K$^+$-ATPase of Chinese hamster ovary cells. Radiat Res 102: 314–323

Arrigo AP (1990) Tumor necrosis factor induces the rapid phosphorylation of the mammalian heat shock protein hsp28. Mol Cell Biol 10: 1276–1280

Arrigo AP, Welch WJ (1987) Purification and characterization of the small 28 kD mammalian stress protein. J Biol Chem 262: 15359–15369

Arrigo AP, Suhan JP, Welch WJ (1988) Dynamic changes in the structure and intracellular locale of the mammalian low-molecular-weight heat shock protein. Mol Cell Biol 8: 5059–5071

Ashburner M, Bonner JJ (1979) The induction of gene activity in Drosophila by heat shock. Cell 17: 241–254

Baulieu EE (1989) J Cell Biochem 35: 161–174

Beckmann RB, Mizzen LA, Welch WJ (1990) Interactions of HSP 70 with newly synthesized proteins: implications for protein folding and assembly. Science 248: 850–854

Bochkaneva ES, Lissin NM, Girshovich AS (1988) Transient association of newly synthesized unfolded proteins with the heat shock GroEL protein. Nature 336: 254–257

Bole DG, Hendershot LM, Kearney JF (1986) Post-translational associations of immunoglobulin heavy chain binding protein with nascent heavy chains in non-secreting and secreting hybridomas. J Cell Biol 102: 1558–1566

Bond U, Schlesinger MJ (1986) Ubiquitin is a heat shock protein in chicken embryo fibroblasts. Mol Cell Biol 5: 949–956

Brugge JS (1989) Interaction of the Rous sarcoma virus proteins pp60src with the cellular proteins pp50 and pp90. Curr Top Microbiol Immunol 123: 1–23

Buffa P, Guarriera-Bobyvela V, Muscatello V, Pasquale I (1970) Conformational changes of mitochondria associated with uncoupling of oxidative phosphorylation in vivo and in vitro. Nature 226: 272

Burdon RH, Cutmore MM (1982) Human heat shock gene expression and the modulation of plasma membrane Na$^+$-K$^+$ ATPase activity. FEBS Lett 140: 45–48

Buzin CH, Bournias-Vardiabasisain N (1982). The induction of a subset of heat shock proteins by drugs that inhibit differentiation in Drosophila embryonic cell cultures. In: Schlesinger M, Ashburner M, Tissiers A (eds) Heat shock: from bacteria to man. Cold Spring Harbor Laboratory, Cold Spring Harbor, NY, pp 387–394

Buzin CH, Bournias-Vardiabasisain N (1984) Proc Natl Acad Sci USA 81: 4075–4079

Calderwood SK, Hahn GM (1983) Thermal sensitivity and resistance of insulin receptor binding. Biochem Biophys Acta 756: 1–8

Caltabiano MM, Koestler TP, Poste G, Greig RG (1986) Induction of 32 and 34 kDa stress proteins by sodium arsenite, heavy metals, and thiol-reactive agents. J Biol Chem 261: 13381–13386

Carlsson L, Lazarodes E (1983) ADP ribosylation of the M_r 83,000 stress-inducible and glucose regulated protein in avian and mammalian cells: modulation by heat shock and glucose starvation. Proc Natl Acad Sci USA 80: 4664–4669

Catelli MG, Binart N, Jung-Testas I, Renoir JM, Baulieu EE, Feramisco JR, Welch WJ (1985) The common 90 kD protein component of non-transformed "8S" steroid receptors is a heat shock protein. EMBO J 4: 3131–3137

Chappell TG, Welch WJ, Schlossman DM, Palter KB, Schlesinger MJ, Rothman JE (1986) Uncoating ATPase is a member of the 70 kDa family of stress proteins. Cell 45: 3–12

Cheney CM, Shearn A (1983) Developmental regulation of Drosophila imaginal disc proteins: synthesis of a heat shock protein under non-heat shock conditions. Dev Biol 95: 325–333

Cheng MY, Ulrich-Hartl F, Martin J, Pollock RA, Kalousek F, Neupert W, Hallberg EM, Hallberg RL, Horwich AL (1989) Mitochondrial heat shock protein hsp 60 is essential for assembly of proteins imported into yeast mitochondria. Nature 337: 620–625

Chirico WJ, Waters MG, Blobel G (1988) 70k heat shock related proteins stimulate protein translocation into microsomes. Nature 333: 805–810

Collier NC, Schlesinger MJ (1986) The dynamic state of heat shock proteins in chicken embryo fibroblasts. J Cell Biol 103: 1495–1507

Coppo A, Manzi A, Pulitzer JF, Takahashi H (1973) Abortive bacteriophage T4 head assembly in mutants of *Escherichia coli*. J Mol Biol 76: 61–87

Craig EA (1985) The heat shock response. CRC Crit Rev Biochem 18: 239–280

Craig E, Jacobsen K (1984) Mutations of the heat inducible 70 kilodalton genes of yeast confer temperature sensitive growth. Cell 38: 841–849

Craig EA, Kramer J, Kosic-Smithers J (1987) SSCI, a member of the 70 kDa heat shock protein multigene family of *Saccharomyces cerevisiae*, is essential for growth. Proc Natl Acad Sci USA 84: 4156–4160

Craig EA, Kramer J, Schilling J, Werner-Washburne M, Holmes S, Kosic-Smithers J, Nicolet CM (1989) SSCI, an essential member of the yeast HSP 70 multigene family, encodes a mitochondrial protein. Mol Cell Biol 9: 3000–3008

Deshaies RJ, Koch BD, Weiner-Washiburne M, Craig E, Schekman R (1988) A subfamily of stress protein facilities translocation of secretory and mitochondrial precursor polypeptides. Nature 333: 800–805

DiDomenico BJ, Bugarsky GE, Lindquist SL (1982) The heat shock response is self-regulated at both the transcriptional and post-transcriptional levels. Cell 31: 593–605

Drummond IA, McClure SA, Poenie M, Tsien RY, Steinhardt RA (1986) Large changes in intracellular pH and calcium observed during heat shock are not responsible for the induction of heat shock proteins in *Drosophila melanogaster*. Mol Cell Biol 6: 1767–1775

Duncan R, Hershey JWB (1984) Heat shock-induced translational alterations in HeLa cells. J Biol Chem 259: 11882–11889

Edington BV, Whelan SA, Hightower LE (1989) Inhibition of heat shock (stress) protein induction by deuterium oxide and glycerol: additional support for the abnormal protein hypothesis of induction. J Cell Physiol 139: 219–228

Edwards DP, Adams DJ, Savage N, McGuire WL (1980) Estrogen induced synthesis of specific proteins in human breast cancer cells. Biochem Biophys Res Commun 93: 804–812

Ellgaard EG, Clever U (1971) RNA metabolism during puff induction in *Drosophila melanogaster*. Chromosome 36: 60–78

Ellgaard EG, Maxwell BL (1975) Nucleotide metabolism in *Drosophila melanogaster* salivary glands during temperature and dinitrophenol-induced puffing. Cell Differ 3: 379–387

Falkner FG, Saumweber H, Biessman H (1981) Two *Drosophila melanogaster* proteins related to intermediate filament proteins of vertebrate cells. J Cell Biol 91: 175–183

Findly RC, Gillies RJ, Shulman RG (1983) In vivo phosphorous-31 nuclear magnetic resonance reveals lowered ATP during heat shock of tetrahymena. Science 219: 1223–1225

Georgopoulos CP, Hohn B (1978) Identification of a host protein necessary for bacteriophage morphogenesis (the *gro E* gene product). Proc Natl Acad Sci USA 75: 131–135

Georgopoulos CP, Hendrix RW, Casjens SR, Kaiser AD (1973) Host participation in bacteriophage lambda head assembly. J Mol Biol 76: 45–60

Gerner EW, Scheider MJ (1975) induced thermal resistance in HeLa cells. Nature 256: 500–502

Gething MJ, McCammon K, Sambrook J (1986) Expression of wild type and mutant forms of influenza hemagglutinin: the role of folding in intracellular transport. Cell 46: 939–950

Goloubinoff P, Gatenby AD, Lorimer GH (1989) GroEL heat shock proteins promote assembly of foreign prokaryotic ribulose bisphosphate carboxylase oligomers in *E. coli*. Nature 337: 44–47

Haas IG, Wabl M (1983) Immunoglobulin heavy chain binding protein. Nature 306: 387–389

Hammond GL, Lai YK, Market CL (1982) Diverse forms of stress lead to new patterns of gene expression through a common and essential metabolic pathway. Proc Natl Acad Sci USA 79: 3485–3488

Heine V, Sverkak L, Kondratuck J, Bonar RA (1971) The behavior of HeLa S3 cells under the influence of supranormal temperatures. J Ultrastruct Res 34: 375–396

Hemmingsen SM, Woolford C, Vandervies SM, Tilly K, Dennie DJ, Georgopoulos C, Hendrix RW, Ellis RJ (1988) Homologous plant and bacterial proteins chaperone oligomeric protein assembly. Nature 333: 330–334

Hendershot L, Bole D, Kohler G, Kearney JF (1987) Assembly and secretion of heavy chains that do not associate post-translationally with immunoglobulin heavy chain binding protein. J Cell Biol 104: 761–767

Henle KJ, Leeper DB (1976) Interaction of hyperthermia and radiation in CHO cells: recovery kinetics. Radiat Res 66: 505–510

Henle KJ, Leeper DB (1982) Modification of the heat response and thermotolerance by cyclohex-imide, hydroxyurea, and lucanthone in CHO cells. Radiat Res 90: 339–347

Hickey ES, Brandom E, Potter R, Stein G, Stein J, Weber LA (1986) Sequence and organization of genes encoding the human 27 kDa heat shock protein. Nucleic Acid Res 14: 4127–4135

Hickey E, Brandon SE, Smale G, Lloyd D, Weber LA (1989) Sequence and regulation of a gene encoding a human 89-kilodalton heat shock protein. Mol Cell Biol 9: 2615–2626

Iida H, Yahara I (1985) Yeast heat shock protein of M_r 48,000 is an isoprotein of enolase. Nature 315: 688–690

Iida K, Iida H, Yahara I (1986) Heat shock induction of intranuclear actin rods in cultured mammalian cells. Exp Cell Res 165: 207–215

Ireland R, Burger E (1982) Synthesis of low molecular weight heat shock proteins stimulated by ecdysterone in a cultured Drosophila cell line. Proc Natl Acad Sci USA 79: 855–859

Itikawa H, Ryu JL (1979) Isolation and characterization of a temperature sensitive DNA K mutant of Escherichia coli. Br J Bacteriol 138: 339–344

Kageyma H, Hiwasa T, Tokunaga K, Sakiyama S (1988) Isolation and characterization of a complementary DNA clone for a Mr32,000 protein which is induced with tumor promoters in Balb/c 3T3 cells. Cancer Res 48: 4795–4798

Kassenbrock CK, Garcia PD, Walter P, Kelly RB (1988) Heavy-chain binding protein recognizes aberrant polypeptides translocated in vitro. Nature 333: 90–93

Kaur P, Welch WJ, Saklatvala J (1989) Interleukin 1 and tumor necrosis factor increase phosphorylation of the small heat shock protein. FEBS Lett 258: 269–273

Keyse SM, Tyrrell RM (1987) Both near ultraviolet radiation and the oxidizing agent hydrogen peroxide induce a 32 kDa stress protein in normal human skin fibroblasts. J Biol Chem 262: 14281–14285

Kim YJ, Shuman T, Sette M, Przybla A (1984) Nuclear localization and phosphorylation of three 25-kilodalton rat stress proteins. Mol Cell Biol 4: 468–473

Kingston RE, Baldwin AS, Sharp PA (1984) Regulation of heat shock protein 70 gene expression by c-myc. Nature 312: 280–283

Koch G, Smith M, Macer D, Webster P, Mortara R (1986) Endoplasmic reticulum contains a common, abundant calcium-binding glycoprotein, endoplasmin. J Cell Sci 86: 217–232

Kost SL, Smith DF, Sullivan WP, Welch WJ, Toft DO (1989) Binding of heat shock proteins to the avian progesterone receptor. Mol Cell Biol 9: 3829–3838

Kozutsumi Y, Segal M, Normington K, Gething MJ, Sambrook J (1988) The presence of malfolded proteins in the endoplasmic reticulum signals the induction of glucose regulated proteins. Nature 332: 462–464

Landry J, Bernier D, Chretien P, Nicole LM, Tanguay RM, Marceau N (1982) Synthesis and degradation of heat shock proteins during development and decay of thermotolerance. Cancer Res 42: 2457–2461

Leenders HJ, Kemp A, Koninkx JF, Rosing J (1974) Changes in cellular ATP, ADP and AMP levels following treatments affecting cellular respiration and the activity of certain nuclear genes in Drosophila salivary glands. Exp Cell Res 86: 25–30

Levinson W, Opperman H, Jackson J (1980) Transition series metals and sulfhydryl reagents induce the synthesis of four proteins in eukaryotic cells. Biochem Biophys Acta 606: 170–180

Lewis MJ, Turco SJ, Green M (1985) Structure and assembly of the endoplasmic reticulum. J Biol Chem 260: 6926–6931

Li CG, Werb Z (1982) Correlation between the synthesis of heat shock proteins and the development of thermotolerance in Chinese hamster fibroblasts. Proc Natl Acad Sci USA 79: 3918–3922

Lindquist S (1986) The heat-shock response. Ann Rev Biochem 55: 1151–1191

Lindquist S, Domenico D, Bugaisky B, Kurtz G, Petko L, Sonoda S (1982) Regulation of the heat shock response in Drosophila and yeast. In: Schlesinger MT, Ashburner M, Tissieres A (eds) Heat shock from bacteria to man. Cold Spring Harbor Laboratory, Cold Spring Harbor, NY, pp 167–175

Magun BE, Fennie CW (1981) Effects of hyperthermia on binding internalization and degradation of epidermal growth factor. Radiat Res 86: 133–146

Marshall JS, DeRocher AE, Keegstra K, Vierling E (1990) Identification of heat shock protein hsp 70 homologues in chloroplasts. Proc Natl Acad Sci USA 87: 374–378

Mayrand S, Pederson T (1983) Heat shock alters ribonucleoprotein assembly in Drosophila cells. Mol Cell Biol 3: 161–171

Mazzarella RA, Green M (1987) ERp 99, an abundant conserved glycoprotein of the endoplasmic reticulum, is homologous to the 90-kDa heat shock protein (hsp90) and the glucose regulated protein (grp 94). J Biol Chem 262: 8875–8883

McCormick P, Keys BJ, Pucci C, Mills AJT (1979) Human fibroblast-conditioned medium contains a 100 k dalton glucose-regulated cell surface protein. Cell 18: 173–192

McMullin TW, Hallberg RL (1987) A normal mitochondrial protein is selectively synthesized and accumulated during heat shock in Tetrahymena thermophila. Mol Cell Biol 7: 4414–4423

Mehdi SQ, Kecktenwald DJ, Smith LM, Li GC, Armour EP, Hahn GM (1984) Effect of hyperthermia on murine cell surface histocompatibility antigens. Cancer Res 44: 3394–3397

Milarski KL, Morimoto R (1986) Expression of human Hsp 70 during the synthetic phase of the cell cycle. Proc Natl Acad Sci USA 83: 9517–9521

Mizzen LA, Welch WJ (1988) Characterization of the thermotolerant cell: I. Effects on protein synthesis activity and the regulation of heat-shock protein 70 expression. J Cell Biol 106: 1105–1116

Mizzen LA, Chang C, Garrels J, Welch WJ (1989) Identification characterization and purification of two mammalian stress proteins present within mitochondria: one related to hsp 70 and the other to the bacterial GroEL protein. J Biol Chem 264: 20664–20675

Mondovi BR, Rotilo G, Argo AF, Cavaliere R, Rossi-Fanelli A (1969) The biochemical mechanism of selective heat sensitivity of cancer cells. I. Studies on cellular respiration. Eur J Cancer 5: 129–136

Moretti-Rojas I, Fugua SA, Montgomery RA, McGuire WL (1988) A cDNA for estradiol-regulated 24k protein: control of mRNA levels in MCF-7 cells. Breast Cancer Res Treat 11: 155–163

Munro S, Pelham HR (1986) An Hsp 70-like protein in the ER: identity with the 78 kd glucose-regulated protein and immunoglobulin heavy chain binding protein. Cell 46: 291–300

Nagata K, Yamada KM (1986) Phosphorylation and transformation sensitivity of a major collagen-binding protein of fibroblasts. J Biol Chem 261: 7531–7536

Nagata K, Saga S, Yamada KM (1986) A major collagen-binding protein of chick embryo fibroblasts is a novel heat shock proteins. J Cell Biol 103: 223–229

Neumann D, Scharf KD, Nover L (1984) Heat shock induced changes of plant cell ultrastructure and autoradiographic localization of heat shock proteins. Eur J Cell Biol 34: 254–264

Nevins JR (1982) Induction of the synthesis of a 70,000 dalton mammalian heat shock protein by the adenovirus E1A gene product. Cell 29: 913–920

Nickells RW, Browder LW (1988) A role of glyceraldehyde-3-phosphate dehydrogenase in the development of thermotolerance in Xenopus laevis embryos. J Cell Biol 107: 1901–1909

Nover L (1984) Heat shock response of eukaryotic cells. Springer, New York Berlin Heidelberg

Olden K, Pratt RM, Yamada KM (1978) Role of carbohydrates in protein secretion and turnover: effects of tunicamycin on the major cell surface glycoprotein of chick embryo fibroblasts. Cell 13: 461–473

Pekkala D, Heath B, Silver JC (1984) Changes in chromatin and the phosphorylation of nuclear protein during heat shock of Achlya ambisexualis. Mol Cell Biol 4: 1198–1205

Pelham HRB (1984) HSP 70 accelerates the recovery of nucleolar morphology after heat shock. EMBO J 3: 3095–3100

Pouyssegur J, Shiu RPC, Pastan I (1977) Induction of two transformation sensitive membrane polypeptides in normal fibroblasts by a block in glycoprotein synthesis or glucose deprivation. Cell 11: 941–947

Rao PN, Engelberg J (1965) HeLa cells: effects of temperature on the life cycle. Science 148: 1092–1094

Rebbe NF, Ware J, Bertina RM, Modrich P, Stafford DW (1987) Nucleotide sequence of a cDNA for a member of the human 90-kDa heat shock proteins family. Gene 53: 235–245

Riabowol KT, Mizzen LA, Welch WJ (1988) Heat shock is lethal to fibroblasts microinjected with antibodies against hsp 70. Science 242: 433–436

Rose DW, Welch WJ, Kramer G, Hardesty B (1989) Possible involvement of the 90-kDa heat shock protein in the regulation of protein synthesis. J Biol Chem 264: 6239–6244

Rubin GM, Hogness DS (1975) Effects of heat shock on low molecular weight RNA's in Drosophila melanogaster: accumulation of a novel form of 5S RNA. Cell 6: 207–213

Sanchez ER, Toft DO, Schlesinger MJ, Pratt WB (1985) Evidence that the 90 kDa phosphoprotein associated with the untransformed L-cell glucocorticoid receptor is a murine heat shock protein. J Biol Chem 260: 12358–12403

Sargan DR, Tsui MJT, O'Malley BW (1986) HSP 108, a novel heat shock inducible protein of chicken. Biochemistry 25: 625–629

Schlossman DM, Schmid SL, Braell WA, Rothman JE (1984) An enzyme that removes clathrin coats: purification of an uncoating ATPase. J Cell Biol 99: 723–733

Schmid SL, Braell WA, Rothman JE (1985) ATP catalyzes the sequestration of clathrin during enzymatic uncoating. J Biol Chem 260: 10057–10062

Simard R, Bernhard W (1967) A heat-sensitive cellular function localized in the nucleolus. J Cell Biol 34: 61–76

Sirotkin K, Davidson N (1982) Developmentally regulated transcription from Drosophila melanogaster chromosomal site. Dev Biol 89: 196–205

Sternberg N (1973) Properties of a mutant of *Escherichia coli* defective in bacteriophage lambda head formation (groEL) II The propagation of phage lambda. J Mol Biol 76: 254

Stevenson MA, Calderwood SK, Hahn GM (1981) Rapid increases in inositol triphosphate and intracellular Ca^{++} after heat shock. Biochem Biophys Res Comm 137: 826–833

Stevenson MA, Minton KW, Hahn GM (1981) Survival and Concanavalin A-induced capping in CHO fibroblasts after exposure to hyperthermia, ethanol and irradiation. Radiat Res 86: 467–478

Subjeck JR, Sciandra J, Johnson RJ (1982) Heat shock proteins and thermotolerance: Comparison of induction kinetics. Br J Radiol 55: 579–584

Subjeck JR, Shyy TT (1986) Stress protein systems of mammalian cells. Am J Physiol 17: 250, C1–C17

Subjeck JR, Shyy T, Shen J, Johnson RJ (1983) Association between mammalian 110k dalton heat shock protein and nucleoli. J Cell Biol 97: 1389–1398

Thomas GP, Welch WJ, Mathews MB, Feramisco JR (1982) Molecular and cellular effects of heat shock and related treatments of mammalian tissue culture cells. Cold Spring Harbor Symp Quant Biol 46: 985–996

Ting J, Lee A (1988) Human gene encoding the 78k dalton glucose regulated protein and its pseudogene: structure, conservation and regulation. DNA 7: 275–286

Ungewickel E (1985) The 70 kDa mammalian heat shock proteins are structurally and functionally related to the uncoating protein that releases clathrin triskelia from coated vesicles. EMBO 4: 3385–3391

Vanbuskirk A, Crump BL, Margoliash E, Pierce SK (1989) A peptide binding protein having a role in antigen presentation is a member of the hsp 70 heat shock family. J Exp Med 170: 1799–1809

Velazquez JM, Lindquist S (1984) HSP 70: nuclear concentration during environmental stress: cytoplasmic storage during recovery. Cell 36: 655–663

Warren AP, James MH, Menzies DE, Widnell CC, PA W, Pasternak CA (1986) Stress induces an increased hexose uptake in cultured cells. J Cell Physiol 128: 383–388

Weinbach EC, Garbus J (1968) Structural changes in mitochondria induced by uncoupling agents. Biochem J 106: 711–747

Weitzel G, Pilatus U, Rensing L (1985) Similar dose response of heat shock protein synthesis and intracellular pH change in yeast. Exp Cell Res 159: 252–256

Welch WJ (1985) Phorbol ester calcium Ionophore, or serum added to quiescent rat embryo fibroblast cells all result in the elevated phosphorylation of two 28,000 dalton mammalian stress proteins. J Biol Chem 260: 3058–3065

Welch WJ, Feramisco JR (1982) Purification of the major mammalian heat shock proteins. J Biol Chem 257: 14949–14959

Welch WJ, Feramisco Jr (1984) Nuclear and nucleolar localization of the 72,000 dalton heat shock protein in heat-shocked mammalian cells. J Biol Chem 259: 4501–4513

Welch WJ, Feramisco JR (1985) Disruption of the three cytoskeletal networks in mammalian cell does not affect transcription, translation, or protein translocation changes induced by heat shock. Mol Cell Biol 5: 1571–1581

Welch WJ, Feramisco JR (1985) Rapid purification of mammalian 70,000 dalton stress proteins: affinity of the proteins for nucleotides. Mol Cell Biol 5: 1229–1237

Welch WJ, Feramisco JR, Blose SH (1985) The mammalian stress response and the cytoskeleton: alterations in the intermediate filaments. Internl. conference on Intermediate Filaments. Ann New York Academy of Sciences 455: 57–67

Welch WJ, Garrels JI, Thomas GP, Lin JJ, Feramisco JR (1983) Biochemical characterization of the mammalian stress proteins and identification of two stress proteins as glucose- and Ca^{2+} ionophore-regulated proteins. J Biol Chem 258: 7102–7111

Welch WJ, Mizzen LA (1988) Characterization of the thermotolerant cell: II. Effects on the intracellular distribution of heat-shock protein 70, intermediate filaments, and small nuclear ribonucleoprotein complexes. J Cell Biol 106: 1117–1130

Welch WJ, Suhan JP (1985) Morphological study of the mammalian stress response characterization of changes in cytoplasmic organelles, cytoskeleton and nucleoli and appearance of intranuclear actin filaments in rat fibroblasts following heat shock treatment. J Cell Biol 101: 1198–1211

Welch WJ, Suhan JP (1986) Cellular and biochemical events in mammalian cells during and after recovery from physiological stress. J Cell Biol 103: 2035–2052

Westra A, Dewey WC (1971) Variation in sensitivity to heat shock during the cell cycle of Chinese hamster ovary cells in vitro. Int J Radiat Biol 19: 467–477

Wu BJ, Hurst AC, Jones NC, Morimoto (1986) The EIA 13S product of adenovirus 5 activates transcription of the cellular human hsp 70 gene. Mol Cell Biol 6: 2994–2999

Yi PN (1979) Cellular ion content changes during and after hyperthermia. Biochem Biophys Res Common 91: 177–182

Yost HJ, Lindquist S (1986) RNA splicing is interrupted by heat shock and is rescued by heat shock protein synthesis. Cell 45: 185–193

Zala CA, Salus-Prato M, Yan WT, Ba jo B, Perdue JF (1980) In cultured chick embryo fibroblasts the hexose transport components are not the 75,000 and 95,000 dalton polypeptides synthesized following glucose deprivation. Can J Biochem 58: 1175–1188

Zantema A, deFong E, Lardenoje K, vanderEb AS (1989) The expression of heat shock proteins hsp 27 and a complexed 22 kilodalton protein is inversely correlated with oncogenicity of Adenovirus-transformed cells. J Vitrol 63: 3368–3375

Zeuthen E (1971) Synchrony in tetrahymena by heat shocks spaced a normal cell generation apart. Ex Cell Res 68: 49–60

Zimmerman JL, Petri W, Meselson M (1971) Accumulation of a specific subset of Drosophila melanogaster heat shock mRNA's in normal development without heat shock. Cell 32: 1161–1168

Heat Shock Protein Genes
and the Major Histocompatibility Complex

E. GÜNTHER

A connection between genes encoding heat shock proteins (hsp's) and genes of
the major histocompatibility complex (MHC) has been detected recently as a
result of immunological and especially of genetic mapping studies. How far the
close genetic linkage observed also implies interactions at the genetic or
functional level between members of both complex gene systems is still unclear,
since pertinent data are scarce.

1 The Structure of the Major Histocompatibility Complex

1.1 Class I and Class II Genes

In many vertebrate species studied so far, notably in mammals, amphibians, and
birds, an MHC has been found (KLEIN 1986; KAUFMAN et al. 1990). Three groups of
MHC genes can be distinguished in mammals: class I, class II, and class III genes.

Abteilung Immungenetik der Universität, Gosslerstrasse 12d, 3400 Göttingen, FRG

Current Topics in Microbiology and Immunology, Vol. 167
© Springer-Verlag Berlin · Heidelberg 1991

The class I and class II genes encode cell surface molecules, which are built of two different pairs of polypeptide chains—heavy chain plus β_2-microglobulin (the latter not being MHC encoded) and α plus β chains—and form a characteristic three-dimensional structure (BJORKMAN et al. 1987). The class I and class II genes and their products are highly polymorphic. The degree of polymorphism and also the number of gene copies in each class may vary between different species. Originally the class I and class II glycoprotein molecules were detected as histocompatibility antigens, since they elicit an immune response which leads to the rejection of transplants from members of the same species. This allograft reaction is the consequence of the large polymorphism of class I and class II molecules and of their potent immunogenicity.

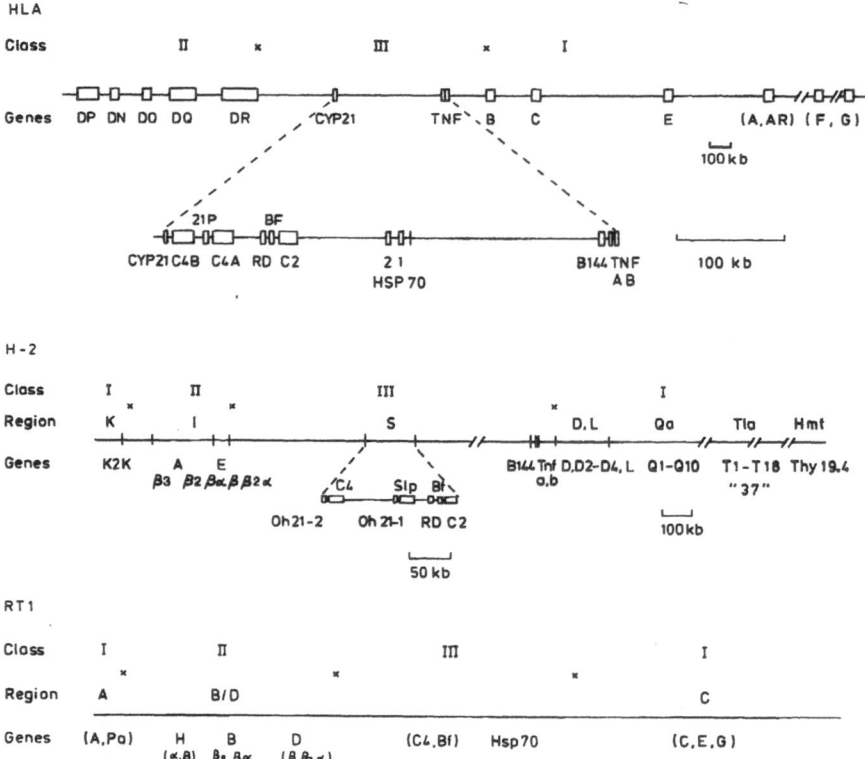

Fig. 1. Schematic representation of the MHCs of man (HLA system), mouse (H-2 system), and rat (RT1 system). The centromer is to the *left* (HLA, H-2 system). The HLA map is mainly based on PFGE and cosmid walking analyses (CARROLL et al. 1987; KOLLER et al. 1989; SARGENT et al. 1989b; SPEISER and WHITE 1989; SPENCE et al. 1989; SPIES et al. 1989). Nomenclature is mostly according to SPENCE et al. (1989). The H-2 map is mainly based on PFGE and cosmid walking data obtained in the BALB/c mouse (STEPHAN et al. 1986; MÜLLER et al. 1987; TSUGE et al. 1987; LEVI-STRAUSS et al. 1988; BRORSON et al. 1989). In general nomenclature is according to GREEN (1989). The RT1 map incorporates data from DIAMOND et al. (1989), GILL et al. (1987), and WURST et al. (1989). Except for the B/D region no physical mapping data are available; grc (not shown) maps to the "right" of E (GILL et al. 1987)

The physiological function of these molecules is to act as accessory receptors for foreign antigen during the immune response, e.g., against infectious agents. They present fragments of the antigen, which has been appropriately processed inside the cell, on the surface of antigen-presenting cells to the antigen receptor on T lymphocytes (restriction function of the class I and class II molecules).

The MHCs of man (HLA), mouse (H-2), and rat (RT1) are the so far best studied mammalian representatives of this genetic system (Fig. 1). The organization of the MHC has been established by classical genetic recombination analysis and by molecular genetic methods resulting in genetic and physical maps according to which the MHC encompasses about 2–4 cM corresponding to $2–4 \times 10^3$ kb. Among the class I multigene family two groups are distinguished: the usually highly polymorphic classical class I genes and the poorly polymorphic class I-like genes; only the products of the former group have been shown to exert a restriction function. Whereas in man a single class I cluster is found, encompassing the *HLA-A,B,C* genes and the *HLA-AR,E,F,G* and further class I-like genes, in mouse and rat the class I genes are separated into two clusters, reflecting the translocation of a few class I genes to the other side of the class II region in a common ancestor of both species. Among the expressed class II genes three groups exist, designated *DP*, *DQ*, and *DR* in man, the latter two being homologous to the expressed mouse class II genes of the *A* and *E* types and the rat *B* and *D* class II genes. Certain domains of the class I and class II genes show homology to each other and to the immunoglobulin genes and thus belong to the immunoglobulin supergene family (WILLIAMS and BARCLAY 1988).

1.2 Class III Genes

The MHC of mammals includes a third cluster, the class III genes, which map between the class II cluster on the one hand and the *HLA-A,B,C, H-2D,Q,T,* and *RT1.C* class I genes on the other (see Fig. 1). On the basis of pulsed-field gel electrophoresis (PFGE) data and cosmid clone analysis a physical map of the class III region has been established in man (CARROLL et al. 1987; SARGENT et al. 1989b; SPIES et al. 1989) and mouse (STEPHAN et al. 1986; MÜLLER et al. 1987). The class III region is about 1000 kb in length and contains so far seven groups of genes known in more detail: the *C4* complement genes, the *C2* and *Bf* complement genes, the 21-hydroxylase genes (*CYP21* in man, *Oh-21* in the mouse), the *RD* gene, the tumor necrosis factor genes (*TNFA,B/Tnfa,b*), the *B144* gene and the *hsp70* genes, which will be discussed in detail below. Further genes have been identified by their transcripts (*BAT1-BAT9*, SPIES et al. 1989; *G1-G11*, SARGENT et al. 1989b). Thus this region is densely populated with expressed genes.

The class III region of the mammalian MHC (WHITE 1989) exhibits several interesting aspects of population and evolutionary genetics. Most genes occur as pairs in man and mouse, reflecting tandem gene duplications. Sequence similarity among the pair members is either extremely high (*C4, CYP21*) or low

(*TNF*). Both members of a pair may usually retain their function (*C4A,B*) or one may have become a pseudogene (*CYP21P/Oh21-2*). Some of the class III genes are very polymorphic, such as *C4A* and *C4B* in man, while others are poorly polymorphic, e.g., the *RD* gene (SPEISER and WHITE 1989).

The homology between pair members favors unequal crossover and gene conversion events, as are observed for the *C4* and *CYP21* genes, resulting in the deletion, amplification, or fusion of the genes (COLLIER et al. 1989). Thus a fraction of the *C4* and *CYP21* deficiencies in man are due to the deletion of the respective gene. As a consequence of unequal crossover and gene conversion, divergence of duplicated genes can be circumvented, a phenomenon which may obscure the phylogenetic analysis of the duplications. Function and tissue specificity of the various known class III gene products are heterogeneous (WHITE 1989). The *C2*, *Bf*, and *C4* genes are expressed in liver and macrophages and are involved in the activation of the complement system, which plays a major effector role in immune responses and inflammation. The *CYP21/Oh21* genes are active in the adrenal cortex and encode 21-hydroxylase, an essential enzyme of steroid hormone synthesis. The *TNFA,B/Tnfa,b* genes are expressed in macrophages and T lymphocytes and code for cytokines. The *B144* gene is transcribed in macrophages and B lymphocytes and is of unknown function (TSUGE et al. 1987). The expression of the *RD* gene (LEVI-STRAUSS et al. 1988) and of the *BAT* and *G* transcripts (SPIES et al. 1989; SARGENT et al. 1989b) on the other hand does not appear to be tissue restricted.

The MHC of *Xenopus* (KAUFMAN et al. 1990) as a representative of amphibians resembles that of mammals with respect to the presence of class I, class II, and class III genes, the latter group being identified by the *C4* gene. In contrast, the chicken MHC is organized differently from that of mammals, and no positive evidence for the presence of genes which are classified as class III genes in mammals has been obtained so far (KROEMER et al. 1990).

2 Heat Shock Protein Genes

The heat shock response—observed after heat shock and many other forms of stress—is phylogenetically very old and the various hsp's, which represent this response but are expressed also under "normal" conditions, are conserved during evolution as well (LINDQUIST and CRAIG 1988). Among the different groups of hsp's, which are distinguished on the basis of their size, those of about 70 kD form the most prominent subset. They constitute a dispersed multigene family and include the major heat shock-responsive *hsp70* gene, a heat shock inducible *hsp72* gene, the constitutively produced and heat-shock enhanced *p72* or *hsc70* gene, the *grp78* gene, which is identical to the gene encoding the heavy chain-binding protein (BiP), and a mitochondrial *hsp75* gene (PELHAM 1986; LINDQUIST and CRAIG 1988; MORIMOTO and MILARSKI 1990). In addition *hsp70*-related transcripts and proteins have been detected in a tissue-specific manner in spermatogenic cells (ALLEN et al. 1988; ZAKERI et al. 1988).

Genes of the *hsp70* family show differential intracellular expression and vary with respect to the inducing stimuli. The *hsp70* gene shows a complex regulation of expression in man; it is basally expressed, cell cycle dependent, and growth regulated, and apart from the common stimulators it is inducible by interleukin-2 (IL-2), serum, and adenovirus or herpesvirus infection (MORIMOTO and MILARKSI 1990).

The function attributed to the hsp70 family in the cell is a protective one and principally related to the interaction with other proteins, hsp70 proteins being involved in the prevention of premature folding or assembly and aggregation, the renaturation of denatured proteins, and the intracellular translocation of proteins.

The members of the hsp90 family do not appear to be structurally related to those of the hsp70 family (LINDQUIST and CRAIG 1988). hsp90 proteins are expressed in the cytoplasm and have been found to be associated with steroid receptors, tyrosine kinases, and certain oncogene products. They appear to be engaged in protein–protein interactions similar to hsp70 proteins. Molecules of 84 and 86 kD acting as tumor-specific transplantation antigens in the mouse have been shown to be homologous to proteins of the hsp90 family (ULLRICH et al. 1986). It is noteworthy that homology also appears to exist between tumor-specific transplantation antigens and molecules of a 100-kD hsp group (SRIVASTAVA et al. 1988).

3 Mapping of Heat Shock Protein Genes to the Major Histocompatibility Complex Chromosome

3.1 *hsp70* Genes

By analyzing somatic cell hybrids for restriction fragment length polymorphisms (RFLPs) with the human *hsp70* probe pH2.3 (Wu et al. 1985), GOATE et al. (1987) were able to map an *hsp70* gene to chromosome 6p21, i.e., to the same position where the HLA complex is localized (SPENCE et al. 1989).

Following the same approach HARRISON et al. (1987) also mapped *hsp70* genes to human chromosome 6 and localized them to 6p21.3-p22. One of the restriction fragments obtained was of the same size as the *hsp70* gene fragment used as probe, i.e., pH2.3, so that the corresponding gene, which had been sequenced before (HUNT and MORIMOTO 1985), most likely is one of the HLA-linked *hsp70* genes. Two additional pH2.3-hybridizing *hsp70* sequences were localized to chromosome 6. All three *hsp70* genes are present in the same 20-kb *Eco*RI fragment and at least one of them is heat shock inducible and thus functional (HARRISON et al. 1987).

On the basis of this information, genetic fine mapping of *hsp70* genes with respect to the various MHC regions was attempted in the rat (WURST et al. 1989). DNA from RT1 recombinant congenic rat strains was analyzed with a human *hsp70* gene probe (DRABENT et al. 1986), representing an *hsp70* gene which is presumably identical to that described by HUNT and MORIMOTO (1985). Under stringent conditions three strongly cross-hybridizing polymorphic *Bam*HI fragments were observed and shown to map within the class III region of the rat MHC (Fig. 1). The same fragments also hybridized with a subclone containing the 5' end of the *hsp* probe, indicating the presence of more than one *hsp70* gene. Only under relaxed stringency did additional fragments become detectable. The latter observation is in accord with the failure of the *hsp70* probe to cross-hybridize with *hsc70* genes under stringent conditions (GOATE et al. 1987).

The analysis of overlapping cosmid clones led SARGENT et al. (1989a) to the identification and precise mapping of two *hsp70* genes, *HSP70-1* and *HSP70-2* (designated *HSPA1* and *HSP1L*, respectively, according to SPENCE et al. 1989), and of an additional *hsp70*-related sequence within the class III region of the human MHC. These *hsp70* genes map close to each other—distances being 11 kb between *HSP70-1* and *HSP70-2* and 5 kb between *HSP70-1* and the related sequence—, about 90 kb telomeric to the *C2* gene and about 280 kb centromeric to the *TNFA* gene. This localization of *HSP70-1* and *HSP70-2* was confirmed by SPIES et al. (1989). Nucleotide sequences corresponding to amino acid residues 369 to 433 are identical between *HSP70-1* and *HSP70-2* (SARGENT et al. 1989a); furthermore, this coding sequence and a partial sequence of the 5' end of the gene are nearly concordant between *HSP70-1* and the sequence published for the major heat shock inducible *hsp70* gene by HUNT and MORIMOTO (1985). Thus the *HSP70-1* gene is identified as the *hsp70* or *hsx70* (PELHAM 1986) gene; this is in accord with the conclusion drawn by HARRISON et al. (1987) from their mapping experiments.

It might be of interest that *hsp70* genes, which cross-hybridize with pH2.3, have been detected in man on at least two other chromosomes, among them chromosome 14 (GOATE et al. 1987; HARRISON et al. 1987), where a functional *hsp70* gene maps to 14q22-q24 between two members of the immunoglobulin supergene family, the immunoglobulin heavy chain locus *IGH* at 14q32.33 and the T cell receptor loci *TCRA* and *TCRD* at 14q11.2.

Preliminary PFGE analysis in the mouse (WURST and GÜNTHER, unpublished data) shows that *hsp70* cross-hybridizing sequences are present on the 530-kb *Mlu*I and the 520-kb *Nru*I fragments, which carry the *Bf* and *C4* genes, but are absent from the 700-kb *Pvu*I fragment carrying the *Bf* and *C4* genes and also absent from the *Tnf*-carrying 120-kb *Pvu*I fragment (see also MÜLLER et al. 1987 and Fig. 1). Thus also in the mouse *hsp70* genes can be mapped into the MHC, where they are found at a position homologous to that described for man and rat. The presence of *hsp70* genes within the MHC can therefore be traced back to the common ancestor of the two lines leading to man and rodents, i.e., for about 70 million years. It will be of interest to study the chicken MHC for the presence of *hsp70* genes in view of the so far unsuccessful attempts to detect class III genes in this avian species.

Since more than one *hsp70* gene is present in the class III region, similar questions concerning expression, polymorphism, and phylogeny of duplication of *hsp70* genes will arise as are discussed above for other class III genes.

The *hsp70* gene identified in the human class III region is expected to reveal certain phenotypic traits, e.g., basal expression, which are not observed in rodents (WELCH and SUHAN 1986; PELHAM 1986). Thus regulation of expression and the promoter region will have to be analyzed separately for each of the MHC-linked *hsp70* genes of the various species.

3.2 *hsp90* Genes

The gene coding for an 84-kD hsp of the hsp90 family has been mapped to the MHC-bearing chromosome 17 of the mouse (MOORE et al. 1987; ROMANO et al. 1989). It is not quite clear whether this *hsp84* gene encodes the tumor-specific transplantation antigen described by ULLRICH et al. (1986) and referred to above. By segregation analysis of interspecies backcross hybrids close linkage of the *hsp84* gene to the MHC was demonstrated and the gene order H-$2E\beta$ ... H-$2L$ ... *hsp84* was suggested on the basis of the recombiration frequencies observed between the H-$2E\beta$ and $Hsp84$ ($r = 7.5$), H-$2L$ and $Hsp84$ ($r = 5.4$), and H-$2E\beta$ and H-$2L$ loci ($r = 2.1$) (ROMANO et al. 1989). Thus in contrast to *hsp70* the *hsp84* gene does not map within but telomeric to the MHC (see also Fig. 1). It should be noted that the recombination frequency between the $E\beta$ and L loci is unusually high compared to that observed in crosses between standard inbred strains.

3.3 Other *hsp*-Related Genes?

The αA- and αB-crystallins of the eye lens show sequence homology to the so-called small hsp's (DE JONG et al. 1988). Interestingly, the gene coding for the αA-crystallin is closely linked to the MHC in the mouse, where it maps to the H-$2K$ side (SKOW and DONNER 1985), and the rat (SKOW et al. 1987) but not in man (HAWKINS et al. 1987).

4 Is the Mapping of Heat Shock Protein Genes to the Major Histocompatibility Complex of Special Significance?

4.1 Genetic and Functional Aspects

No structural and consequently no phylogenetic relationship has been found between the class I and class II MHC genes on the one hand and the known class III genes on the other hand, nor between the various groups of class III genes. The clustering of the class III genes and their localization within the MHC could thus be due to chance, the class III region representing "entrapped loci" and "frozen blocks" (KLEIN 1986; KLEIN and TAKAHATA 1990). Similarly, the localization of the *hsp84* and αA-crystallin genes close to the MHC could indicate nothing other than that some genes have to be neighbors.

An alternative explanation implies that the MHC genes or their products interact with each other in a manner that natural selection favors close linkage of them. In particular, the fact that certain allelic combinations (haplotypes) of the various MHC genes—including the *C4, C2, Bf,* and *CYP21* genes of the class III region—occur more frequently in the population than is to be expected from the frequencies of the single alleles, has been attributed to natural selection in favor of such linkage disequilibria.

Evidence is accumulating which shows at the DNA level that homologous regions of different MHC haplotypes need not be of identical length (COLLIER et al. 1989; DUNHAM et al. 1989; WHITE 1989), thus preventing recombination due to misalignment, and that recombination frequenceis are not evenly distributed within the MHC (SHIROISHI et al. 1990). This might be taken as an argument against the involvement of natural selection in maintaining linkage equilibria; however, the distribution of recombination sites could be affected by natural selection as well.

It is not difficult to present examples for connections between MHC and *hsp70* genes or their products. Thus a motif resembling the sequence of the heat shock element has been detected in the promoter of H-2 class I genes (KIMURA et al. 1986). The expression of class I genes is suppressed (VAESSEN et al. 1986) and the *hsp70* gene is activated by the adenovirus E1A protein (MORIMOTO and MILARKSI 1990). *hsp70* is inducible by IL-2, which on the other hand regulates in an autocrine manner the proliferation of MHC-restricted T lymphocytes. TNF-α is able to destroy cells, which become more resistant against this attack after heat shock (GROMKOWSKI et al. 1989). Obvious as such connections may be, it is unclear how far they indicate the necessity of close linkage between the genes involved.

One, might, nevertheless, speculate that the presence of so many expressed genes in close proximity, as is the case with the class III region, and possibly the MHC as a whole, could be meaningful beyond the level of single genes, especially since members of different defense or protection systems—specific immunity, complement system, inflammation, heat shock response—meet in the MHC. It could have turned out during evolution that this assembly of genes is favorable for their appropriate function in protecting the organism.

Considering the interaction of hsp's with newly synthesized or abnormal proteins accumulating in the cell, the mapping of *hsp70* genes inside the MHC sheds new light on three important MHC-related issues of immunogenetics: the intracellular traffic of class I and class II polypeptides, the processing of antigens, and the association between MHC and disease susceptibility. One could imagine that hsp70 is involved in the processing of the endogenous class I-restricted antigens. Evidence has been reported for the participation of hsp70-like proteins in the presentation of class II-restricted antigens (VANBUSKIRK et al. 1989) and for the presentation of hsp70 molecules or hsp70 fragments on the surface of cells (for review see KAUFMANN 1990). It is, however, unclear whether the hsp's involved are products of MHC-linked *hsp70* genes.

4.2 HLA–Disease Associations

The mapping of *hsp* genes to the MHC provides new markers for studying associations between the HLA system and certain diseases, which often are of an autoimmune nature (TIWARI and TERASAKI 1985). Thus the genetic analysis will be refined, but also new concepts are introduced for investigating the pathogenesis of some of the MHC-associated diseases. Only in certain instances has susceptibility to the disease been assigned to distinct HLA genes. It is of great importance to identify those HLA genes which in reality are the susceptibility-conferring factors—especially with respect to the linkage disequilibria—and to differentiate between single genes and haplotype constellations as being responsible for a given association.

Interest will focus on diseases, susceptibility to which appears to be primarily associated with class III region genes. Examples are systemic lupus erythematosus (SLE), IgA deficiency, and certain types of common variable immunodeficiency. The latter two syndromes appear to be caused by a defect not of the immunoglobulin genes but of the differentiation of B lymphocytes to antibody-secreting cells. Both diseases are more frequent in individuals carrying HLA haplotypes with certain rare *C4* and *C2* alleles, *C4* null alleles, and deletions of the *CYP21* pseudogene (SCHAFFER et al. 1989). Susceptibility to SLE is found closely associated with deficiencies of the *C4* and *C2* complement genes. Furthermore, among the multitude of autoantibodies occurring in SLE patients antibodies to hsp90 (MINOTA et al. 1988a) and the constitutively expressed hsp70 (MINOTA et al. 1988b) are also observed. It is speculated that anti-hsp70 antibodies could interfere with the nucleolus-protecting role described for hsp70 (LINDQUIST and CRAIG 1988). Antibodies to the heat-inducible hsp70 have been found in some patients with infectious mononucleosis (MINOTA et al. 1988b). Antibody activity against human hsp70 above that detectable in normal sera has been observed in rheumatoid arthritis, ankylosing spondylitis (two HLA-associated diseases), and tuberculosis (TSOULFA et al. 1989). In the course of many bacterial and protozoal infections antibodies and T lymphocytes reactive with hsp's of the infectious agent are produced (for review see KAUFMANN 1990). Because of the homology between protozoan, bacterial, and autologous hsp's and the homology between the various human hsp70 molecules, epitope sharing is assumed to contribute to the induction of autoimmune reactions in the course of anti-hsp immunity. Since the immune reaction to antigens s controlled by the MHC class I and class II molecules, MHC-associated high and low responsive-ness could be expected for anti-hsp immune responses. No association was found between anti-hsp65 antibody levels and HLA class II phenotype (BAHR et al. 1988).

Polymorphism of MHC-linked *hsp* genes, possibly differential organization of the *hsp* genes in the class III region as a consequence of unequal crossover, but also polymorphism exhibited by the ligands of the hsp's, could contribute to interindividual variability in susceptibility to certain diseases, hsp70 variants could turn out to be of particular relevance under extreme conditions of cell function and during chronic pathological processes. The initial approach to look for associations between *hsp* genes and disease susceptibility will make use of RFLPs. In the rat RFLPs have been detected in several MHC haplotypes and with 2 of the 13 restriction enzymes tested (WURST et al. 1989). In man an RFLP has been described using the pH2.3 probe and the restriction enzyme *PstI*, which is HLA-linked according to family studies (GOATE et al. 1987) and thus indeed affects MHC-linked *hsp70* gene(s). The two alleles, represented by a 10- or an 8.5-kb fragment, occur with frequencies of 0.62 and 0.38, respectively, the polymorphism information content value being relatively low (0.24). In a further study (BENESCH and GÜNTHER, unpublished data) a similar polymorphism was detected (8.9-kb and 7.6-kb fragments, allele frequencies being 0.57 and 0.43, respectively).

Work is now in progress to characterize the *hsp70* genes of the class III region in detail, which together with a better knowledge of the other so far ill-defined class III region genes will soon answer several of the questions raised above and left open at present.

Acknowledgments. The author would like to thank Drs. R.D. Campbell and R.I. Morimoto for unpublished information and the DFG for support of his work cited in this review.

References

Allen RL, O'Brien DA, Eddy EM (1988) A novel hsp 70-like protein (P70) is present in mouse spermatogenic cells. Mol Cell Biol 8: 828–832

Bahr GM, Rook GAW, Al-Saffar M, Van Embden J, Stanford JL, Behbehani K (1988) Antibody levels to mycobacteria in relation to HLA type: evidence for non-HLA-linked high levels of antibody to the 65 kD heat shock protein of *M. bovis* in rheumatoid arthritis. Clin Exp Immunol 74: 211–215.

Bjorkman PJ, Saper MA, Samraoui B, Bennett WS, Strominger JL, Wiley DC (1987) Structure of human class I histocompatibility antigen, HLA-A2. Nature 329: 506–512

Brorson KA, Richards S, Hunt III SW, Cheroutre H, Fischer Lindahl K (1989) Analysis of a new class I gene mapping to the Hmt region of the mouse. Immunogenetics 30: 273–283

Carroll MC, Katzman P, Alicot EM, Koller BH, Geraghty DE, Orr HT, Strominger JL, Spies T (1987) Linkage map of the human major histocompatibility complex including the tumor necrosis factor genes. Proc Natl Acad Sci USA 84: 8535–8539

Collier S, Sinnott PJ, Dyer PA, Price DA, Harris R, Strachan T (1989) Pulsed field gel electrophoresis identifies a high degree of variability in the number of tandem 21-hydroxylase and complement C4 gene repeats in 21-hydroxylase deficiency haplotypes. EMBO J 8: 1393–1402

De Jong WW, Hendriks W, Mulders JWM, Bloemendal H (1989) Evolution of eye lens crystallins: the stress connection. TIBS 14: 365–368

Diamond AG, Hood LE, Howard JC, Windle M, Winoto A (1989) The class II genes of the rat MHC. J Immunol 142: 3268–3274

Drabent B, Genthe A, Benecke B-J (1986) In vitro transcription of a human hsp 70 heat shock gene by extracts prepared from heat-shocked and non-heat-shocked human cells. Nucleic Acids Res 14: 8933–8948

Dunham I, Sargent CA, Dawkins RL, Campbell RD (1989) An analysis of variation in the long-range genomic organization of the human major histocompatibility complex class II region by pulsed-field gel electrophoresis. Genomics 5: 787–796

Gill TJ III, Kunz HW, Misra DN, Cortese Hassett AL (1987) The major histocompatibility complex of the rat. Transplantation 43: 773–785

Goate AM, Cooper DN, Hall C, Leung TKC, Solomon E, Lim L (1987) Localization of a human heat-shock HSP70 gene sequence to chromosome 6 and detection of two other loci by somatic-cell hybrid and restriction fragment length polymorphism analysis. Hum Genet 75: 123–128

Green MC (1989) Catalog of mutant genes and polymorphic loci. In: Lyon MF, Searle AG (eds) Genetic variants and strains of the laboratory mouse, 2nd edn, pp 12–403. Oxford University Press, Oxford; Gustav Fischer, Stuttgart

Gromkowski SH, Yagi J, Janeway CA Jr (1989) Elevated temperature regulates tumor necrosis factor-mediated immune killing. Eur J Immunol 19: 1709–1714

Harrison GS, Drabkin HA, Kao F-T, Hartz J, Hart IM, Chu EHY, Wu BJ, Morimoto RI (1987) Chromosomal location of human genes encoding major heat-shock protein HSP70. Somatic Cell Mol Genet 13: 119–130

Hawkins JW, Van Keuren ML, Piatigorsky J, Law ML, Patterson D, Kao F-T (1987) Confirmation of assignment of the human α1-crystallin gene (CRYA1) to chromosome 21 with regional localization to q22.3. Hum Genet 76: 375–380

Hunt C, Morimoto RI (1985) Conserved features of eukaryotic HSP70 genes revealed by comparison with the nucleotide sequence of human HSP70. Proc Natl Acad Sci USA 82: 6455–6459

Kaufman J, Skoedt K, Salomonsen J (1990) The MHC molecules of nonmammalian vertebrates. Immunol Rev 113: 83–117

Kaufmann SHE (1990) Heat shock proteins and the immune response. Immunol Today 11: 129–136

Kimura A, Israel A, Le Bail O, Kourilsky P (1986) Detailed analysis of the mouse H-2K^b promotor: enhancer-like sequences and their role in the regulation of class I gene expression. Cell 44: 261–272

Klein J (1986) Natural history of the major histocompatibility complex. John Wiley, New York

Klein J, Takahata N (1990) The major histocompatibility complex and the quest for origins. Immunol Rev 113: 5–25

Koller BH, Geraghty DE, DeMars R, Duvick L, Rich SS, Orr HT (1989) Chromosomal organization of the human major histocompatibility complex class I gene family. J Exp Med 169: 469–480

Kroemer G, Bernot A, Behar G, Chausse A-M, Gastinel L-N, Guillemot F, Park , Thoraval P, Zoorob R, Auffray C (1990) Molecular genetics of the chicken MHC: current status and evolutionary aspects. Immunol Rev 113: 119–145

Levi-Strauss M, Carroll MC, Steinmetz M, Meo T (1988) A previously undetected MHC gene with an unusual periodic structure. Science 240: 201–204

Lindquist S, Craig EA (1988) The heat-shock proteins. Annu Rev Genet 22 631–677

Minota S, Koyasu S, Yahara I, Winfield JB (1988a) Autoantibodies to the heat-shock protein hsp90 in systemic lupus erythematosus. J Clin Invest 81: 106–112

Minota S, Cameron B, Welch WJ, Winfield JB (1988b) Autoantibodies to the constitutive 73-kD member of the HSP70 family of heat shock proteins in systemic lupus erythematosus. J Exp Med 168: 1475–1480

Moore SK, Kozak C, Robinson EA, Ullrich SJ, Appella E (1987) Cloning and nucleotide sequence of the murine hsp84 cDNA and chromosome assignment of related sequences. Gene 56: 29–40

Morimoto RI, Milarski KL (1990) Expression and function of vertebrate hsp70 genes. In: Morimoto RI, Tissières A, Georgopoulos C (eds) Stress proteins in biology and medicine. Cold Spring Harbor Laboratory, Cold Spring Harbor NY

Müller U, Stephan D, Philippsen P, Steinmetz M (1987) Orientation and molecular map position of the complement genes in the mouse MHC. EMBO J 6: 369–373

Pelham HRB (1986) Speculations on the functions of the major heat shock and glucose-regulated proteins. Cell 46: 959–961

Romano JW, Seldin MF, Appella E (1989) Linkage of the mouse Hsp84 heat shock protein structural gene to the H-2 complex. Immunogenetics 29: 142–144

Sargent CA, Dunham I, Trowsdale J, Campbell RD (1989a) Human major histocompatibility complex contains genes for the major heat shock protein HSP70. Proc Natl Acad Sci USA 86: 1968–1972

Sargent CA, Dunham I, Campbell RD (1989b) Identification of multiple HTF-island associated genes in the human major histocompatibility complex class III region. EMBO J 8: 2305–2312

Schaffer FM, Palermos J, Zhu ZB, Barger BO, Cooper MD, Volanakis JE (1989) Individual with IgA deficiency and common variable immunodeficiency share polymorphisms of major histocompatibility complex class III genes. Proc Natl Acad Sci USA 86: 8015–8019.

Shiroishi T, Hanzawa N, Sagai T, Ishiura M, Gojobori T, Steinmetz M (1990) Recombinational hotspot specific to female meiosis in the mouse major histocompatibility complex. Immunogenetics 31: 79–88

Skow LC, Donner ME (1985) The locus encoding αA-crystallin is closely linked to H-2K on mouse chromosome 17. Genetics 110: 723–732

Skow LC, Kunz HW, Gill TJ III (1985) Linkage of the locus encoding the A chain of α-crystallin (Acry-1) to the major histocompatibility complex in the rat. Immunogenetics 22: 291–293

Speiser PW, White PC (1989) Structure of the human RD gene: a highly conserved gene in the class III region of the major histocompatibility complex. DNA 8: 745–751

Spence MA, Spurr NK, Field LL (1989) Report of the committee on the genetic constitution of chromosome 6. Cytogenet Cell Genet 51: 149–165

Spies T, Bresnahan M, Strominger JL (1989) Human major histocompatibility complex contains a minimum of 19 genes between the complement cluster and HLA-B. Proc Natl Acad Sci USA 86: 8955–8958

Srivastava PK, Kozak CA, Old LJ (1988) Chromosomal assignment of the gene encoding the mouse tumor rejection antigen gp90. Immunogenetics 28: 205–207

Stephan D, Sun H, Fischer Lindahl K, Meyer E, Hämmerling G, Hood L, Steinmetz M (1986) Organization and evolution of D region class I genes in the mouse major histocompatibility complex. J Exp Med 163: 1227–1244

Tiwari JL, Terasaki PI (1985) HLA and disease associations. Springer, New York Berlin Heidelberg Tokyo

Tsoulfa G, Rook GAW, Bahr GM, Sattar MA, Behbehani K, Young DB, Mehlert A, Van Embden JDA, Hay FC, Isenberg DA, Lydyard PM (1989) Elevated IgG antibody levels to the mycobacterial 65-kDa heat shock protein are characteristic of patients with rheumatoid arthritis. Scand J Immunol 30: 519–527

Tsuge I, Shen F-W, Steinmetz M, Boyse EA (1987) A gene in the H-2S:H-2D interval of the major histocompatibility complex which is transcribed in B cells and macrophages. Immunogenetics 26: 378–380

Ullrich SJ, Robinson EA, Law LW, Willingham M (1986) A mouse tumor-specific transplantation antigen is a heat shock-related protein. Proc Natl Acad Sci USA 83: 3121–3125

Vaessen RTMJ, Houweling A, Israel A, Kourilsky P, van der Eb AJ (1986) Adenovirus E1A-mediated regulation of class I MHC expression. EMBO J 5: 335–341

VanBuskirk A, Crump BL, Margoliash E, Pierce SK (1989) A peptide binding protein having a role in antigen presentation is a member of the HSP70 heat shock family. J Exp Med 170: 1799–1809

Welch WJ, Suhan JP (1986) Cellular and biochemical events in mammalian cells during and after recovery from physiological stress. J Cell Biol 103: 2035–2052

White PC (1989) Molecular genetics of the class III region of the HLA complex. In: Dupont B (ed) Immunobiology of HLA, vol II. Springer, Berlin Heidelberg New York, pp 62–69

Williams AF, Barclay AN (1988) The immunoglobulin superfamily—domains for cell surface recognition. Ann Rev Immunol 6: 381–405

Wu B, Hunt C, Morimoto RI (1985) Structure and expression of the human gene encoding major heat shock protein HSP70. Mol Cell Biol 5: 330–341

Wurst W, Benesch C, Drabent B, Rothermel E, Benecke B-J, Günther E (1989) Localization of heat shock protein 70 genes inside the rat major histocompatibility complex close to class III genes. Immunogenetics 30: 46–49

Zakeri ZF, Ponzetto C, Wolgemuth DJ (1988) Translational regulation of the novel haploid-specific transcripts for the c-abl proto-oncogene and a member of the 70 kDa heat-shock protein gene family in the male germ line. Dev Biol 125: 417–422

Note Added in Proof

In the goat the class III region of the MHC is reported to be similarly organized as in man and mouse and to include hsp70 genes at a homologous position (Cameron et al. Immunogenetics 31: 253–264, 1990). The BAT3 gene mapping into the HLA class III region is shown to contain an ubiquitin-like domain and heat shock elements (Banerji et al. Proc. Natl. Acad. Sci. U.S.A. 87: 2374–2378, 1990).

Heat Shock Protein
Functions Related to Immunity

BiP—A Heat Shock Protein Involved in Immunoglobulin Chain Assembly

I. G. HAAS

1 Introduction

1.1 The Experimental System

Precursor cells of B lymphocytes (pre-B cells) arise from bone marrow stem cells and are characterized by the intracellular expression of immunoglobulin (Ig) heavy (H) chains in the absence of light (L) chains. Later in ontogeny, B lymphocytes also synthesize L chains and express the assembled Ig molecule on the surface. Driven by antigenic stimulation, these cells then differentiate into plasma cells which secrete high amounts of antibodies.

The hybridoma technology has made possible the creation of permanent cells lines reflecting the various differentiation stages with respect to the Ig pattern they express (e.g., hybridomas derived from pre-B cells synthesize Ig H chain only, which is neither secreted nor expressed on the surface). Because of their biochemical and molecular homogeneity, hybridomas have been a useful tool to study the process of Ig chain assembly. Most of the analyses presented below have been performed with such hybridoma lines.

Institut für Genetik, Universität zu Köln, Zülpicher Str. 47, 5000 Köln, FRG

Current Topics in Microbiology and Immunology, Vol. 167
© Springer-Verlag Berlin · Heidelberg 1991

1.2 The Discovery of BiP

Specific immunoprecipitations of Ig H chains from cell lines lacking L chain expression revealed the appearance of an additional molecule apparently not related to Ig (MORRISON and SCHARFF 1975; BURROWS et al. 1981). We demonstrated that these immunoprecipitates contained a specific 78-kD protein which interacts noncovalently with Ig H chains in pre-B and plasma cells but only if the H chains are not associated with L chains. Furthermore, the formation of the complex is independent of the isotype of the H chain. For these reasons, we named this molecule Ig H chain binding protein (BiP) (HAAS and WABL 1983). Cell fractionation experiments enriched BiP together with the endoplasmic reticulum (ER) although some of the protein could readily be isolated from the postmicrosomal supernatant, suggesting that BiP is a soluble component of the ER (BOLE et al. 1986). Molecular cDNA cloning of BiP revealed that the protein is related to the heat shock protein (hsp) 70 family and that BiP is indeed an ER protein (MUNRO and PELHAM 1986; HAAS and MEO 1988).

2 General Characteristics of BiP

2.1 What are Heat Shock Proteins?

The original definition of hsp's concerned exclusively those proteins which are expressed in response to physiological stress. They are categorized according to their molecular weight. Besides a heterogeneous group of small hsp's, there is an hsp60, an hsp70 (the most highly conserved family), and an hsp90 (also called hsp83) family. It has been shown that the constitutive expression of hsp70 accelerates the recovery from heat shock induced changes in the nucleolar morphology of transfected monkey COS cells (PELHAM 1984). Meanwhile, a number of proteins have been identified which are indeed expressed in a constitutive fashion and which are structurally related to hsp's. These are referred to as heat shock cognates (hsc's) (for review, see LINDQUIST and CRAIG 1988).

A feature common to probably all hsp70 members is their ability to bind ATP very tightly (WELCH and FERAMISCO 1985). Recently, evidence has accumulated suggesting that hsp molecules are involved in ATP dependent processes like protein translocation [e.g., into microsomes (CHIRICO et al. 1988) or into mitochondria (DESHAIES et al. 1988)] and protein folding [e.g., in mitochondria (OSTERMANN et al. 1989) or in bacteria (GOLOUBINOFF et al. 1989)]. The new term "chaperonin" has been introduced particularly for those proteins which assist in folding and/or assembly of oligomeric protein structures (ELLIS 1987; HEMMINGSEN et al. 1988). A number of recent reviews deal with these issues

(ROTHMAN and KORNBERG 1986; PELHAM 1988, 1989; ROTHMANN 1989; SAMBROOK and GETHING 1989).

2.2 Structure, Expression, and Intracellular Localization of BiP

Primary structure (cDNA cloning) and serological analyses have revealed that BiP is not only a member of the hsp70 family but that it is also identical to the glucose regulated protein (grp)78 (MUNRO and PELHAM 1986; HAAS and MEO 1988; HENDERSHOT et al. 1988) which was originally described in transformed chick embryo fibroblasts (SHIU et al. 1977). The typical characteristic of grp's is that their synthesis is enhanced in cells which are deprived of glucose (for review, see LEE 1987). There are two well-characterized grp's, grp78 (BiP) and grp94, which is an hsc90 protein (MAZZARELLA and GREEN 1987; SMITH and KOCH 1987; SORGER and PELHAM 1987).

BiP is ubiquitously expressed though different amounts are found in different tissues (HAAS, unpublished data). The protein is widely distributed among vertebrates and has been cloned from mouse (HAAS and MEO 1988), rat (MUNRO and PELHAM 1986), hamster (TING et al. 1987), chicken (STOECKLE et al. 1988), and man (TING and LEE 1988). Moreover, BiP analogues have been found in nematodes (77% amino acid sequence homology to rat BiP, HESCHL and BAILLIE 1989), plasmodium (72%, MATTEI et al. 1988), and yeast (67% NORMINGTON et al. 1989; ROSE et al. 1989; NICHOLSON et al. 1990). The overall expression and the extremely high conservation in the amino acid sequence of BiP—two of the four amino acid exchanges observed between hamster and rat are within the signal peptide and thus are not present in the mature protein (TING et al. 1987)—not only suggests a common ancestor molecule but also indicates that the protein has a very essential function. Indeed, KAR-2, which is the BiP homologue in yeast, is essential for cell growth. Furthermore, the presence of mouse BiP can complement a defective allele of KAR-2 (NORMINGTON et al. 1989; ROSE et al. 1989).

Like other soluble residents of the ER such as protein disulfate isomerase (EDMAN et al. 1985) and grp94, BiP has a carboxyterminal signal sequence, KDEL, which is responsible for its retention in the ER, since the removal of this sequence leads to secretion of BiP (MUNRO and PELHAM 1987). Interestingly, the perturbation of cellular calcium levels by the calcium ionophore A23187 also induces the secretion of BiP and other luminal ER proteins despite the presence of the KDEL sequence (BOOTH and KOCH 1989). This effect has been shown not to be due to a general overexpression of ER residents overloading the KDEL retention system. BOOTH and KOCH propose that ER residents form higher ordered structures and thus build an immobile matrix which is stabilized by calcium ions. This view is supported by the finding that BiP and other ER residents exhibit calcium binding properties (MACER and KOCH 1988). Furthermore, members of the hsp70 family contain a highly conserved calmodulin binding domain (STEVENSON and CALDERWOOD 1990). In this model, true secretory proteins would be part of a mobile phase in the ER. Thus, an association with BiP might be sufficient to prevent them from being secreted.

2.3 BiP–Ligand Interaction and Regulation

The original discovery of BiP was due to its property of binding H chains if those are not associated to L chains (HAAS and WABL 1983). Meanwhile, a number of other BiP ligands have been described. These include polypeptides as different as a folding mutant of viral hemagglutinin (HA) (GETHING et al. 1986), unglycosylated forms of factor VIII, von Willebrand factor, and tissue plasminogen activator (DORNER et al. 1987), incompletely disulfide bonded prolactin, and incompletely glycosylated invertase (KASSENBROCK et al. 1988). However, it is not clear in every case whether all or only a portion of the respective ligand molecules are associated with BiP in the cell. This is important for the functional conclusions which can be drawn from these results. Furthermore, it would also be interesting to determine the molar ratio of the components in the BiP–ligand complex.

A feature common to most of the structures that have been found in association with BiP seems to be a malfolding of the proteins due to an incorrect primary structure, incomplete disulfide bond formation, incomplete glycosylation, or incomplete subunit assembly. This poses the question of whether each of these polypeptides possesses a special BiP binding site or whether there is a structural feature common to all of them defining the ability to bind to BiP. One might imagine that hydrophobic regions normally hidden away inside the structure would be exposed on the surface of malfolded polypeptides. In this way, the interaction of malfolded molecules with BiP would be driven by nonspecific hydrophobic interactions. Though this may be part of the answer, the observation that short hydrophilic synthetic peptides also can associate with BiP (FLYNN et al. 1989) suggests that the situation may be more complex. Thus, the real basis of BiP–ligand interaction is far from understood.

It appears that the accumulation of malfolded proteins inside the ER can induce an increase in the steady state level of BiP. The most direct evidence for this has been provided by J. SAMBROOK and his collaborators. Using a viral expression vector, they introduced a modified form of a viral HA, which was incapable of trimerization, into simian CV-1 cells. The mutant protein not only associates with BiP but also causes an increase in the steady state levels of BiP mRNA and protein (KOZUTSUMI et al. 1988). This result suggests that the expression may be regulated by the level of BiP which is available for ligand interaction. An analogous situation is observed in yeast, where hsp70 proteins have been reported to regulate their own synthesis (CRAIG and JACOBSEN 1984; STONE and CRAIG 1990). Although in a less direct way, the other treatments inducing BiP expression could also be explained by an accumulation of malfolded polypeptides in the ER: increased BiP levels are observed where glycosylation is inhibited. This can be achieved either by depleting cells of glucose (POUYSSEGUR et al. 1977; LIN and LEE 1984) or by the addition of glycosylation inhibitors such as tunicamycin or glucosamine (WATOWICH and MORIMOTO 1988). Astonishingly, β-mercaptoethanol (KIM and LEE 1987) and the calcium ionophore A23187 (RESENDEZ et al. 1985; DRUMMOND et al. 1987; MACER and KOCH 1988) also have stimulating effects on BiP mRNA expression. However, the latter effects may similarly be due to a block in glycosylation, since an inverse correlation of glycosylation and steady state level of BiP transcripts has been described (CHANG et al. 1987).

An interesting finding has been that the regulation of BiP can also occur at the level of mRNA translation. Infection of human HeLa cells with poliovirus leads to increased translation of unaltered levels of BiP mRNA at a time when cap dependent translation of other cellular mRNAs is greatly reduced (SARNOW 1989).

Irrespective of whether regulation is at the level of transcription, translation, or both, it is clear that a signal has to be transferred from inside the ER to the cell nucleus or another relevant compartment in order to establish a feedback regulation mechanism. In addition, the cell must be capable of discriminating free and ligand occupied forms of BiP.

BiP is an ATP binding protein and the BiP/H chain complex is dissociated in the presence of ATP but not in the presence of non-hydrolyzable ATP analogues (MUNRO and PELHAM 1986). This has also been shown for other hsp70–ligand complexes (CHAPELL et al. 1986; OSTERMANN et al. 1989). The dissociation is probably driven by the ATPase activity attributed to BiP (KASSENBROCK and KELLY 1989). The weak ATPase activity which BiP displays on its own is enhanced in the ligand bound form (FLYNN et al. 1989). In this context, it is interesting to note that BiP phosphorylation could be detected in the free but not in the H chain associated form (HENDERSHOT et al. 1988). Thus, it is possible that BiP is dephosphorylated in the H chain association process and/or that autophosphorylation takes places while BiP disassembles from H chains. Autophosphorylation activity has been reported for the E. coli dnaK protein, which is also a member of the hsp70 family (ZYLICZ et al. 1983). It is conceivable that protein phosphorylation provides the cell with a means of measuring the levels of free and ligand bound forms of BiP.

Of all tissues and cell lines analyzed so far, the highest levels of BiP are seen in pre-B cell derived hybridomas which also synthesize high levels of Ig H chains in the absence of L chains. In hybridoma lines, high levels of free H chains always seem to require high levels of BiP, whereas chain loss often correlates with a decreased expression of BiP (HAAS, unpublished results). Furthermore, Abelson virus transformed pre-B cells which produce very little H chain protein (JÄCK and WABL 1988) also express only low amounts of BiP (MUNRO and PELHAM 1986). Thus, the interaction of BiP with H chains resembles the interaction of BiP with other ligands: the more BiP is occupied by ligand interaction, the more BiP is produced by the cell. Because of this relationship, it is very likely that the various BiP levels expressed in the different tissues reflect the actual amount of ligand available to interact with BiP.

This may also explain why BiP is found in a high molar excess compared to the total amount of H chains in pre-B cells. Because of its ubiquitous nature, BiP is certain not only to bind to Ig H chains. On the contrary, it is rather likely that a lot of additional molecules destined for export also associate with BiP in pre-B cells. However, no defined polypeptide other than H chain has yet been identified. Since further putative BiP ligands probably represent a heterogeneous population of molecules which, in addition, may bind only transiently to BiP, it would be technically difficult to detect them. One approach to identify other BiP ligands would be to overexpress one component of a multisubunit complex in the absence of its appropriate partner subunit(s) and, by so doing, mimic the arrest of an otherwise transient event.

3 BiP and Immunoglobulins

3.1 The Domain Structure of Immunoglobulins

A typical immunoglobulin consists of two identical H and two identical L chains covalently linked to each other by disulfide bonds. The various Ig classes and subclasses like IgM, IgD, IgG_1, IgG_{2a}, IgG_{2b}, IgG_3, IgA, and IgE are defined by the type of H chain (isotype) which is present. Both H and L chains are composed of domains, each of which is about 110 amino acids in length and exhibits a common pattern of polypeptide chain folding stabilized by an internal disulfide bond. The amino terminal domains are variable in both H and L chains (VH and VL respectively). The antigen binding site is defined by the combination of paired V region domains in the assembled Ig molecule. The other domains are constant and define the respective chain isotype. The H chains of the δ, γ, and α isotypes have three constant domains, while those of the μ and ϵ isotypes have four (CH1-4). The most C-terminal constant domain of the H chain determines whether the Ig molecule will be secreted or inserted into the membrane, a phenomenon which is dependent on differential mRNA splicing. Disulfide bonds in the secretory tailpieces are responsible for the formation of pentamers in IgM or of dimers in IgA. All other immunoglobulins are monomeric in their membrane as well as in their secreted forms. Depending on the isotype, the two H chains are linked by a variable number of disulfide bonds which are usually determined by a short peptide sequence between CH1 and CH2. The L chains have only one constant domain (CL), which is involved in the interchain linkage to the CH1 domain of the H chain (for more details on Ig structure, see BURTON 1987).

3.2 BiP and Heavy Chains

The association of H chains with BiP has been described for various H chain isotypes, e.g., mouse μ, $\gamma 2b$, $\gamma 3$ (HAAS and WABL 1983), mouse μ, $\gamma 1$, (BOLE et al. 1986), mouse μ, $\gamma 1$, $\gamma 2b$, α (HENDERSHOT et al. 1987), and the secretory as well as the membrane form of human μ chain (HENDERSHOT and KEARNEY 1988). This indicates that the binding to BiP is a common property of all H chain isotypes. Since BiP binding is abolished once L chains are bound to H chains (HAAS and WABL 1983; BOLE et al. 1986), it is possible that BiP binding involves the constant domain of H chains which is also responsible for L chain association. This idea is supported by the results of an analysis of mutant H chains: whereas deletion of CH2 or CH3 domains does not influence BiP binding, deletion of CH1 leads to the secretion of free H chains and BiP binding becomes undetectable (MORRISON 1978; HENDERSHOT et al. 1987). Furthermore, H chains secreted in human H chain disease lack L chains and have deletions which, in most cases, include the CH1 domain (reviewed in SELIGMANN et al. 1979).

However, there are exceptions which do not fit this picture. Obviously, deletions in the V region of human μ chains can be responsible for membrane expression of these H chains in the absence of L chains in B lymphoid cell lines (POLLOK et al. 1987). In addition, an H chain disease protein has been described which has a deletion in the V region but still has an apparently intact CH1 domain (FRANKLIN et al. 1976). Furthermore, a chimeric molecule consisting of human CD4 linked to the constant region of human γ1 chains is secreted in the absence of L chains, even though this molecule still has a complete CH1 domain (CAPON et al. 1989). Taken together, it appears that the CH1 domain is required for BiP association but can be influenced by V region structures or, more generally, by sequences N-terminal to the CH1 domain. What remains valid to date is that H chains deleted for CH1 do not interact with BiP unless an alternative specific BiP binding site is present.

In B cells, the secretory tailpiece of μ has been shown to be responsible for the retention of CH1 deleted μ chains. In addition, the same tailpiece causes the retention of IgG molecules which contain a chimeric H chain (SITIA et al. 1990). Thus, deletion of both the CH1 and the secretory tailpiece is required to obtain secretion of free μ H chains. This might explain why only about 5% of the H chain disease proteins are of the μ isotype.

Overall, these findings support the idea that BiP binding involves those portions of the H chains which are also implicated in normal chain assembly or oligomerization of the Ig molecule.

With regard to the functional significance of the BiP/H chain complex it is important to consider the fate of BiP bound H chains. In pre-B cells, all H chains are bound to BiP and have to be degraded since they are not exported from the cell. [The degradation pathway of H chains is as yet unknown; some recent data propose a novel, nonlysosomal pre-Golgi compartment for the degradation of ER residents (BONIFACINO et al. 1989).] In contrast, H chains associate only transiently with BiP in cells which also produce L chains (BOLE et al. 1986). By the combination of pulse chase labeling and cell fusion techniques, it has been possible to demonstrate that those H chains which are bound to BiP in H chain only producing cells can be assembled and secreted as complete Ig molecules after cell fusion to L chain producing cells (HENDERSHOT 1990). From this result, it is evident that H chains are able to disassemble from BiP and to participate in the formation of functional Ig molecules.

3.3 BiP and Light Chains

Myelomas or hybridomas which lose Ig H expression continue to synthesize and secrete their L chains (BAUMAL et al. 1973). In some cases, however, the L chains are not exported but remain intracellular. In one of these cell lines, the L chain monomers have been reported to associate with BiP (NAKAKI et al. 1989). Surprisingly, these L chains are still able to assemble with H chains (COLMAN et al. 1982), suggesting that a malfolding caused by a fundamentally aberrant structure cannot be the reason why BiP binds to this particular L chain. Thus, it is tempting to speculate that BiP also binds to normal L chains, an event which might not readily be seen in hybridoma cells because BiP/L chain complexes would disintegrate when L chains dimerize or assemble with H chains.

3.4 A Postulated Role for BiP in Immunoglobulin Chain Assembly

In contrast to the ability of plasma cells to express L chains in the absence of H chains, the loss of L chain expression while H chains are still present is apparently lethal (KÖHLER 1980; WILDE and MILSTEIN 1980), except when he cells are mutagenized or express a mutant H chain (MORRISON and SCHARFF 1975; MORRISON 1978; SONENSHEIN et al. 1978). This phenomenon, also called H chain toxicity, could be explained by the tendency of free H chains to form large, insoluble aggregates which would occlude the ER and thus damage the cells. In contrast to this phenomenon, pre-B cells are perfectly able to synthesize H chains in the absence of L chains. Interestingly, this pre-B cell property can be transferred to plasma cells by cell fusion (HAAS and WABL 1984). As mentioned above, hybridomas derived from pre-B cells show very high levels of BiP. We proposed that BiP might play a critical role in neutralizing H chain toxicity by preventing H chains from aggregating. Thus, the simplest way to explain the difference in the ability of pre-B and plasma cells to express H chains in the absence of L chains is that the amount of BiP expressed in plasma cells is too low to counteract a sudden loss of L chain expression.

   An important contribution to the understanding of BiP function is provided by the finding that BiP bound H chains are not necessarily degraded but are capable of participating in the formation of functional Ig molecules (see above). This suggests that BiP is functionally involved in the process of Ig chain assembly. There are two different ways to view the phenomenon of BiP–H chain complexes in the context of H/L chain assembly. First, it might be that BiP binds only to those H chains which, for whatever reason, gain a malfolded conformation and thus are susceptible to aggregation. These H chain could then either enter the degradation pathway (e.g., in pre-B cells) or, by dissociation from BiP, be given the chance to refold into their native conformation and reenter the chain assembly pathway (in B or plasma cells). In this model, BiP performs only a recycling or salvage activity. Alternatively, BiP may associate with all H chains and function to catalyze the Ig assembly process if L chains are available.

However, for what could a catalyzing activity in the assembly of H and L chains be required? In the case of mitochondrial hsp60, it has been shown that the association causes the ligand to retain its unfolded state. Only after disassembly from hsp60 does the molecule fold into its mature conformation (OSTERMANN et al. 1989). If this finding is transferred to the situation in Igs, it seems possible that native H chains also might not acquire an assembly competent conformation on their own. In this context, it is conceivable that the CH1 domain has to be unfolded in order to interact and assemble efficiently with L chains. Therefore, BiP association with H chains when the intrachain disulfide bond is not yet formed in CH1 might serve to keep the CH1 in an unfolded conformation (similarly, unfolding of the constant region may also be required on the L chain side). Chain assembly could be initiated by interaction of the VH and VL region domains, which also might provide the signal for BiP dissociation. Disulfide bonds could then be formed, a process which is probably driven by protein disulfide isomerase (ROTH and PIERCE 1987), and the molecule could be exported from the ER. In this model, the complex of H chain and BiP found in pre-B cells would reflect a normal step in the Ig assembly pathway which, however, is arrested because of missing L chains.

Experiments performed in *Xenopus* oocytes, which allow the synthesis of H and L chains at separate time points, revealed that Ig chain assembly can occur as a posttranslational event (COLMAN et al. 1982). This is compatible with the view that (oocyte) Bip is involved in the chain association process. Alternatively, Ig chains may also assemble cotranslationally: completed L chains associate with nascent H chains which have a minimal length of 38 kD (BERGMANN and KUEHL 1979). This means that V region, CH1, and CH2 have already formed before L chains can interact. Thus, if BiP was also to catalyze the cotranslational assembly of H and L chains, the prediction would be that B P associates with nascent H chains. In any case, this would be the easiest way in which BiP could bind to unfolded CH1 domains.

In principle, the postulated role of BiP in Ig assembly can easily be transferred to the interaction of BiP with other ligands. Bip might principally bind to that portion of the ligand which has to retain an unfolded conformation in order to associate with either another portion of the molecule in monomeric proteins or another subunit in multiunit proteins. According to this view, BiP would not associate with malfolded molecules per se but would act to retain polypeptide chains until all posttranslational modifications required for folding or assembly have been accomplished. Nevertheless, the phenomenon of BiP/H chains complexes remains a natural model system appropriated for studying the physiological role of BiP in protein folding and assembly processes.

Acknowledgments. I greatly appreciated the helpful comments of R. Jack and A. Starzinski-Powitz, who read the first version of this manuscript through the eyes of nonBiPologists. Furthermore, I would like to thank K. Rajewsky for his critical reading of the final version and S. Vogel for her help in typing.

References

Baumal R, Birshtein BK, Coffino P, Scharff MD (1973) Mutations in immunoglobulin-producing mouse myeloma cells. Science 182: 164–166

Bergmann LW, Kuehl WM (1979) Formation of intermolecular disulfide bonds on nascent immunoglobulin polypeptides. J Biol Chem 254: 5690–5694

Bole DG, Hendershot LM, Kearney JF (1986) Posttranslational association of immunoglobulin heavy chain binding protein with nascent heavy chains in nonsecreting and secreting hybridomas. J cell Biol 102: 1558–1566

Bonifacino JS, Suzuki CK, Lippincott-Schwartz J, Weissman AL, Klausner RD (1989) Pre-Golgi degradation of newly synthesized T cell antigen receptor chains: intrinsic sensitivity and the role of subunit assembly. J Cell Biol 109: 73–83

Booth C, Koch GLE (1989) Perturbation of cellular calcium induces secretion of luminal ER proteins. Cell 59: 729–737

Burrows PD, Beck GB, Wabl MR (1981) Expression of μ and γ immunoglobulin heavy chains in different cells of a cloned mouse lymphoid line. Proc Natl Acad Sci USA 78: 564–568

Burton DR (1987) Structure and function of antibodies. In: Calabi F, Neuberger MS (eds) Molecular genetics of immunoglobulins. Elsevier, Amsterdam, pp 1–50

Capon DJ, Chamov SM, Mordenti J, Marsters SA, Gregory T, Mitsuya H, Byrn RA, Lucas C, Wurm FM, Groopman JE, Broder S, Smith DH (1989) Designing CD4 immunoadhesins for AIDS therapy. Nature 337: 525–531

Chang SC, Wooden SK, Nakaki T, Kim YK, Lin AY, Kung L, Attenello JW, Lee AS (1987) Rat gene encoding the 78 kDa glucose-regulated protein GRP78: its regulatory sequences and the effect of protein glycosylation on its expression. Proc Natl Acad Sci USA 84: 680–684

Chappell TG, Welch WJ, Schlossman DM, Palter KB, Schlesinger MJ, Rothman JE (1986) Uncoating ATPase is a member of the 70 kilodalton family of stress proteins. Cell 45: 3–13

Chirico WJ, Waters MG, Blobel G (1988) 70K heat shock related proteins stimulate protein translocation into microsomes. Nature 332: 805–810

Colman A, Besley J, Valle G (1982) Interactions of mouse immunoglobulin chains within Xenopus oocytes. J Mol Biol 160: 459–474

Craig EA, Jacobsen K (1984) Mutations of the heat inducible 70 kilodalton genes of yeast confer temperature sensitive growth. Cell 38: 841–849

Deshaies RJ, Koch BD, Werner-Washburne M, Craig EA, Schekman R (1988) A subfamily of stress proteins facilitates translocation of secretory and mitochondrial precursor polypeptides. Nature 332: 800–805

Dorner AJ, Bole DG, Kaufman RJ (1987) The relationship of N-linked glycosylation and heavy chain-binding protein association with the secretion of glycoproteins. J Cell Biol 105: 2665–2674

Drummond AS, Lee AS, Resendez E, Steinhardt RA (1987) Depletion of intracellular calcium stores by calcium ionophore A23187 induces the genes for glucose-regulated proteins in hamster fibroblasts. J Biol Chem 262: 12801–12805

Edman JC, Ellis L, Blacher RW, Roth RA, Rutter WJ (1985) Sequence of protein disulphide isomerase and implications of its relationship to thioredoxin. Nature 317: 267–270

Ellis J (1987) Proteins as molecular chaperones. Nature 328: 378–379

Flynn GC, Chappell TG, Rothman JE (1989) Peptide binding and release by proteins implicated as catalysts of protein assembly. Science 245: 385–390

Franklin EC, Frangione B, Prelli F (1976) The defect in μ heavy chain disease protein GLI. J Immunol 116: 1194–1195

Gething MJ, McCammon K, Sambrook J (1986) Expression of wild-type and mutant forms of influenza hemagglutinin: the role of folding in intracellular transport. Cell 46: 939–950

Goloubinoff P, Gatenby AA, Lorimer GH (1989) GroE heat-shock proteins promote assembly of foreign prokaryotic ribulose bisphosphate carboxylase oligomers in Escherichia coli. Nature 337: 44–47

Haas IG, Meo T (1988) cDNA cloning of the immunoglobulin heavy chain binding protein. Proc Natl Acad Sci USA 85: 2250–2254

Haas IG, Wabl M (1983) Immunoglobulin heavy chain binding protein. Nature 306: 387–389

Haas IG, Wabl M (1984) Immunoglobulin heavy chain toxicity in plasma cells is neutralized by fusion to pre-B cells. Proc Natl Acad Sci USA 81: 7185–7188

Hemmingsen SM, Woolford C, van der Vies SM, Tilly K, Dennis DT, Georgopoulos CP, Hendrix RW, Ellis RJ (1988) Homologous plant and bacterial proteins chaperone oligomeric protein assembly. Nature 333: 330–334

Hendershot LM (1990) Immunoglobulin heavy chain and binding protein complexes are dissociated in vivo by light chain addition. J Cell Biol (in press)

Hendershot LM, Kearney JF (1988) A role for human heavy chain binding protein in the developmental regulation of immunoglobulin transport. Mol Immunol 224: 585–595

Hendershot LM, Bole D, Köhler G, Kearney JF (1987) Assembly and secretion of heavy chains that do not assemble post translationally with immunoglobulin heavy chain binding protein. J Cell Biol 104: 761–767

Hendershot LM, Ting J, Lee AS (1988) Identity of the immunoglobulin heavy chain binding protein with the 78,000-dalton glucose-regulated protein and the role of posttranslational modifications in its binding function. Mol Cell Biol 8: 4250–4256

Heschl MFP, Baillie DL (1989) Characterization of the hsp70 multigene family of Caenorhabditis elegans. DNA 8: 233–243

Jäck HM, Wabl M (1988) Immunoglobulin mRNA stability varies during B lymphocyte differentiation. EMBO J 7: 1041–1046

Kassenbrock CK, Kelly RB (1989) Interaction of heavy chain binding protein (BiP/GRP78) with adenine nucleotides. EMBO J 8: 1461–1467

Kassenbrock CK, Garcia PD, Walter P, Kelly RB (1988) Heavy chain binding protein recognizes aberrant polypeptides translocated in vitro. Nature 333: 90–93

Kim YK, Lee AS (1987) Transcriptional activation of the glucose-regulated protein genes and their heterologous fusion genes by β-mercaptoethanol. Mol Cell Biol 7: 2974–2976

Köhler G (1980) Immunoglobulin chain loss in hybridoma lines. Proc Natl Acad Sci USA 77: 2197–2199

Kozutsumi Y, Segal M, Normington K, Gething MJ, Sambrook J (1988) The presence of malfolded proteins in the endoplasmic reticulum signals the induction of glucose-regulated proteins. Nature 332: 462–464

Lee AS (1987) Coordinated regulation of a set of genes by glucose and calcium ionophores in mammalian cells. TIBS 12: 20–23

Lin AY, Lee AS (1984) Induction of two genes by glucose starvation in hamster fibroblasts. Proc Natl Acad Sci USA 81: 988–992

Lindquist S, Craig EA (1988) The heat-shock proteins. Ann Rev Genet 22: 631–677

Macer DPJ, Koch GLE (1988) Identification of a set of calcium-binding proteins in reticuloplasm, the luminal content of the endoplasmic reticulum. J Cell Sci 92: 61–70

Mattei D, Ozaki LS, Pereira da Silva L (1988) A Plasmodium falciparum gene encoding a heat shock-like antigen to the rat 78 kDa glucose regulated protein. Nucleic Acids Res 16: 5204

Mazzarella RA, Green M (1987) ERp99, an abundant, conserved glycoprotein of the endoplasmic reticulum, is homologous to the 90-kDa heat shock protein (hsp90) and the 94-kDa glucose regulated protein (GRP94). J Biol Chem 262: 8875–8883

Morrison SL (1978) Murine heavy chain disease. Eur J Immunol 8: 194–199

Morrison SL, Scharff MD (1975) Heavy chain producing variants of a mouse myeloma cell line. J Immunol 114: 655–659

Munro S, Pelham HRB (1986) An hsp70-like protein in the ER: identity with the 78 kd glucose-regulated protein and immunoglobulin heavy chain binding protein. Cell 46: 291–300

Munro S, Pelham HRB (1987) A C-terminal signal prevents secretion of luminal ER proteins. Cell 48: 899–907

Nakaki T, Deans RJ, Lee AS (1989) Enhanced transcription of the 78,000 dalton glucose-regulated protein (GRP78) gene and association of GRP78 with immunoglobulin light chains in a nonsecreting B-cell myeloma line (NS-1). Mol Cell Biol 9: 2233–2238

Nicholson RC, Williams DB, Moran LA (1990) An essential member of the HSP70 gene family of Saccharomyces cerevisiae is homologous to immunoglobulin heavy chain binding protein. Proc Natl Acad Sci USA 86: 1159–1163

Normington K, Kohno K, Kozutsumi Y, Gething MJ, Sambrook J (1989) S. cerevisiae encodes an essential protein homologous in sequence and function to mammalian BiP. Cell 57: 1223–1236

Ostermann J, Horwich AL, Neupert W, Hartl FU (1989) Protein folding in mitochondria requires complex formation with hsp60 and ATP hydrolysis. Nature 341: 125–130

Pelham HRB (1984) Hsp70 accelerates the recovery of nucleolar morphology after heat shock. EMBO J 3: 3095–3100

Pelham HRB (1988) Heat shock proteins: coming in from the cold. Nature 332: 776–777

Pelham HRB (1989) Heat shock and the sorting of luminal ER proteins. EMBO J 8: 3171–3176

Pollok BA, Anker R, Eldridge P, Hendershot LM, Levitt D (1987) Molecular basis of the cell-surface expression of immunoglobulin μ chain without light chain in human B lymphocytes. Proc Natl Acad Sci USA 84: 9199–9203

Pouyssegur J, Shiu RPC, Pastan I (1977) Induction of two transformation sensitive membrane polypeptides in normal fibroblasts by a block in glycoprotein synthesis or glucose deprivation. Cell 11: 941–947

Resendez E, Attenello JW, Grafsky A, Chang CS, Lee AS (1985) Calcium ionophore A23187 induces expression of glucose-regulated genes and their heterologous fusion genes. Mol Cell Biol 5: 1212–1219

Rose MD, Misra LM, Vogel JP (1989) KAR2, a karyogamy gene, is the yeast homolog of the mammalian BiP/GRP78 gene. Cell 57: 1211–1221

Roth RA, Pierce SB (1987) In vivo cross-linking of protein disulfide isomerase to immunoglobulins. Biochemistry 26: 4179–4182

Rothman JE, Kornberg RD (1986) An unfolding story of protein translocation. Nature 322: 209–210

Rothmann JE (1989) Polypeptide chain binding proteins: catalysts of protein folding and related processes in cells. Cell 59: 591–601

Sambrook J, Gething MJ (1989) Chaperones, paperones. Nature 342: 224–225

Sarnow P (1989) Translation of glucose-regulated protein 78/immunoglobulin heavy-chain binding protein mRNA is increased in poliovirus-infected cells at a time when cap-dependent translation of cellular mRNAs is inhibited. Proc Natl Acad Sci USA 86: 5795–5799

Seligmann M, Mihaesco E, Preud'homme JL, Danon F, Brouet JC (1979) Heavy chain diseases: current findings and concepts. Immunol Rev 48: 145–167

Shiu RPC, Pouyssegur J, Pastan I (1977) Glucose depletion accounts for the induction of two transformation-sensitive membrane proteins in Rous sarcoma virus-transformed chick embryo fibroblasts. Proc Natl Acad Sci USA 74: 3840–3844

Sitia R, Neuberger M, Alberini C, Bet P, Fra A, Valetti C, Williams G. Milstein C (1990) Developmental regulation of IgM secretion: the role of the carboxy-terminal cysteine. Cell 60: 781–790

Smith MJ, Koch GLE (1987) Isolation and identification of partial cDNA clones for endoplasmin, the major glycoprotein of mammalian endoplasmic reticulum. J Mol Biol 194: 345–347

Sonenshein GE, Siekevitz M, Siebert GR, Gefter ML (1978) Control of immunoglobulin secretion in the murine plasmacytoma line MOPC 315. J Exp Med 148: 301–312

Sorger PK, Pelham HRB (1987) The glucose-regulated protein grp94 is related to heat shock protein hsp90. J Mol Biol 194: 341–344

Stevenson MA, Calderwood SK (1990) Members of the 70-kilodalton heat shock protein family contain a highly conserved calmodulin-binding domain. Mol Cell Biol 108: 1234–1238

Stoeckle MY, Sugano S, Hampe A, Vashistha A, Pellman D, Hanafusa H (1988) 78-kilodalton glucose-regulated protein is induced in Rous sarcoma virus-transformed cells independently of glucose deprivation. Mol Cell Biol 8: 2675–2680

Stone DE, Craig EA (1990) Self-regulation of 70-kilodalton heat shock proteins in Saccharomyces cerevisiae. Mol Cell Biol 10: 1622–1632

Ting J, Lee AS (1988) Human gene encoding the 78,000 dalton glucose-regulated protein and its pseudogene: structure, conservation, and regulation. DNA 7: 275–286

Ting J, Wooden SK, Kriz R, Kelleher K, Kaufman RJ, Lee AS (1987) The nucleotide sequence encoding the hamster 78-kDa glucose-regulated protein (GRP78) and its conservation between hamster and rat. Gene 55: 147–152

Watowich SS, Morimoto RI (1988) Complex regulation of heat shock- and glucose-responsive genes in human cells. Mol Cell Biol 8: 393–405

Welch WJ, Feramisco JR (1985) Rapid purification of mammalian 70,000-dalton stress proteins: affinity of the proteins for nucleotides. Mol Cell Biol 5: 1229–1237

Wilde CD, Milstein C (1980) Analysis of immunoglobulin chain secretion using hybrid myelomas. Eur J Immunol 10: 462–467

Zylicz M, LeBowitz JH, McMacken R, Georgopoulos C (1983) The dnak protein of Escherichia coli possesses an ATPase and autophosphorylating activity and is essential in an in vitro DNA replication system. Proc Natl Acad Sci USA 80: 6431–6435

A Role for Heat Shock Proteins in Antigen Processing and Presentation

S. K. PIERCE,[1] D. C. DeNAGEL,[1] and A. M. VanBUSKIRK[2]

1 An Overview of Antigen Processing and Presentation

T lymphocytes recognize antigen through their cell surface immune receptor, the T cell receptor (TCR). In several respects the TCR is similar to the B cell immune receptor, antibody. Both are composed of two different protein chains which associate to form a combining site for antigen. These chains have variable and constant regions encoded in separate genes which rearrange during the development of the individual cells. Indeed, TCR and antibody are members of the same super gene family and based on amino acid sequence similarity the TCR is predicted to have a similar three-dimensional structure as antibody (reviewed in DAVIS and BJORKMAN 1988). Despite such similarities the T cell and B cell recognition of antigen differs fundamentally. Antibody binds to protein antigens in their native conformation in an exquisitely specific, lock and key fashion, as revealed by recent determinations of the crystal structures of antibodies bound to protein antigens (AMIT et al. 1986). In contrast, the TCR has no apparent affinity for native protein antigens but requires that the antigen be denatured, most likely by proteolysis, yielding a peptide or peptide-like fragment containing the T cell antigenic determinant. In general, antigenic peptides alone do not bind to TCRs but require an association with the MHC molecule on the surface of an antigen-presenting cell (APC) to activate antigen-specific T cells.

[1] Department of Biochemistry, Molecular Biology and Cell Biology, Northwestern University, Evanston, IL 60208, USA
[2] Department of Obstetrics and Gynecology, College of Medicine, The Ohio State University, Columbus, Ohio 43210, USA

Current Topics in Microbiology and Immunology, Vol. 167
© Springer-Verlag Berlin · Heidelberg 1991

This appears to be the case both for helper T cells which recognize antigen in association with MHC class II molecules (Ia) and for cytolytic T cells which recognize antigen bound to MHC class I molecules. T cells express one of two types of TCR. THe vast majority of mature adult T cells express a TCR composed of α and β chains while a minority (5%–10%) express a receptor composed of γ and δ chains (reviewed in MATIS 1990). It is well documented that T cells expressing $\alpha\beta$ TCR recognize antigenic peptides when presented by MHC molecules of the class I or class II type. However, at present it is controversial whether $\gamma\delta$ T cells recognize antigen in a similar fashion and whether antigenic peptides are presented on the MHC class I or II molecules for $\gamma\delta$ T cell recognition. With regard to the role of heat shock proteins in immune responses, the nature of the antigen recognized by $\gamma\delta$ T cells is of considerable interest and is addressed by others in this volume. Here we will restrict our discussion to the T cell recognition of antigenic peptides bound to MHC molecules and in particular to antigen presented by Ia and the potential role for heat shock proteins in this process. Although the overlap in mechanisms by which antigen is processed and presented via class I and class II molecules is not known, we would leave open the possibility of similar roles for the heat shock proteins in both pathways.

The molecular mechanisms by which native antigens are converted to peptides and presented by the MHC molecules are, at present, only poorly delineated. For Ia-recognizing T cells, antigen processing is a function of Ia-expressing APCs. Processing involves the internalization of antigen by the APC, transport of the antigen to an acidic compartment, degradation or proteolysis of the antigen, and then subsequent association of the resulting peptide fragments with the Ia molecule (HARDING et al. 1988). All Ia-expressing cells appear competent to process and present antigen, including cells which do not normally express, Ia, for example fibroblasts transfected with the genes encoding the Ia molecule (MALISSEN et al. 1984). Thus, antigen processing would not appear to be a specialized function of the differentiated cells which express Ia molecules, the most common being B cells and macrophages. Rather, the minimal machinery necessary to process antigen and to present it with Ia would appear to be present in most, if not all cell types. Whether cells which normally express Ia have specialized mechanisms which make antigen processing and presentation more efficient remains to be determined.

Antigen can enter APCs through fluid phase pinocytosis, by phagocytosis, or in the cases where APCs express a receptor for antigen, as for antigen-specific B cells, by receptor-mediated endocytosis (LANZAVECCHIA 1990). The receptor-mediated pathway for antigen processing is extremely efficient as compared to pinocytosis. Indeed, specific B cells activate T cells when provided with 1/1000th-1/10 000th the antigen concentration necessary for nonspecific B cells. Antigen receptors on B cells may not be the only APC surface structures capable of enhancing antigen processing. In macrophages, the Fc receptor may function to internalize antigen-containing immune complexes or the phagosome itself may represent a specialized means of antigen internaliztion. At present it is controversial whether proteins which are synthesized intracellularly have access to the Ia processing pathway and under which conditions such a pathway might function. Studies of the processing of influenza viral proteins indicate that intracellularly synthesized viral proteins cannot be presented with Ia even though such proteins can be processed and presented with Ia when the protein is provided as killed virus (MORRISON et al. 1986). In contrast, studies with measles virus (JACOBSON et al. 1989) and B cell antibody (WEISS and BOGEN 1989) acting as an antigen indicate that intracellularly synthesized antigen is readily presented with Ia. This is an important issue for which further studies are necessary to resolve the controversy.

The intracellular compartment(s) in which processing occurs and where processed antigen associates with Ia has not been identified. Whether newly synthesized Ia and/or Ia recycling into the cell from the surface associates with processed antigen has not been resolved. BRODSKY and co-workers (GUAGLIARDI et al. 1990) have recently described a potential intracellular site for antigen processing. Using immunoelectron microscopy, they demonstrated the endosomal co-localization of Ig, Ia, and the proteases cathepsin B and D in a human lymphoblastoid B cell line which had been allowed to internalize anti-immunoglobulin for 2 min. While these results offer a good candidate for a processing compartment, no data currently link this endosomal compartment with the site of processing. Studies by CRESSWELL (1990) indicate that newly synthesized Ia cannot bind peptide until released of its accompanying invariant chain. This release is believed to occur in a post-Golgi compartment immediately prior to transport to the plasma membrane. Ploegh and co-workers (NEEFJES et al. 1990) have shown that newly synthesized Ia is retained for 2 h in a post-Golgi endosomal compartment which has access to pinocytosed material and have suggested this is a potential site of Ia–antigenic peptide association. The compartment in which processing occurs is of significant interest but its identification will require further studies.

Presumably the final step in the processing pathway involves the binding of an antigenic peptide to the MHC molecule. The binding of peptides representing T cell antigenic determinants has been demonstrated in vitro. Initial studies were carried out with purified Ia in detergent solution (BABBITT et al. 1985; BUUS et al. 1986) and more recently with purified Ia incorporated into synthetic membranes (SADEGH-NASSERI and MCCONNELL 1989). These studies revealed highly unusual characteristics of the peptide–Ia interaction. Firstly, only a small fraction of the total Ia isolated from cells was able to bind peptide. This portion ranged from 1 % to 10 %. Although there are several possible explanations for this observation, one frequently cited is that the majority of the Ia purified from a cell contains antigenic peptides resulting from the processing of either endogenous antigens or exogenous proteins from the cell's environment. Indeed, the crystal structure of an MHC class I molecule showed unresolvable electron density in a groove between two α-helices (BJORKMAN et al. 1987). This was presumed to represent peptides which remained bound to the MHC molecules through rigorous purification procedures. Recent studies from our laboratory (SRINIVASAN and PIERCE 1990) show that Ia molecules purified from cells which have processed a foreign antigen do indeed contain antigenic peptides, as shown by the ability of the complex, when incorporated into a synthetic membrane, to stimulate specific T cells. This finding indicates that an antigenic peptide resulting from processing, once associated with Ia, remains bound in a nearly irreversible fashion. Indeed, kinetic analysis of the binding of synthetic peptides to purified Ia supports a long-lived complex. These studies showed an average equilibrium constant of approximately 10^{-6} M with an extraordinary slow association ($t_{1/2}$ = 5 h) and dissociation rate ($t_{1/2}$ = 30 h) (BUUS et al. 1986). Furthermore, recent studies by WATTS and co-workers indicate that extreme conditions are required to release synthetic peptide bound to purified Ia (pH 2 for several hours) (LEE and WATTS 1990).

Such observations from in vitro systems are somewhat puzzling in that they would not be predicted from the biology of antigen processing and presentation. APCs can rapidly associate with synthetic peptides within minutes, resulting in functional complexes capable of stimulating specific T cells (LAKEY et al. 1988; ROOSNEK et al. 1988). Once peptides associate with APCs and free peptide is washed away the rate of loss of complexes, as measured by loss of the ability to stimulate T cells, is also relatively rapid, being complete within 4 h. The observed loss of peptide from the cell surface cannot be accounted for by turnover of Ia. Thus, the APCs would appear to have the capability of assembling and disassembling peptide–Ia complexes more rapidly than the rates measured in vitro would indicate. One may need to be cautious in comparing the binding of peptides to purified Ia and to APCs because peptide association with Ia on cell surfaces can only be measured by the formation of a stimulatory complex and the subsequent activation of a T cell. Thus, the interpretation of the results with APCs is clouded by the vagaries of the requirements for T cell activation in terms of the number of peptide–Ia complexes necessary to activate the T cell and the potential role of the T cell in facilitating or stabilizing the peptide–Ia interactions. Nevertheless, there would seem to be a discrepancy in the observations studying peptide binding to APCs and to purified Ia. Although there may be several explanations for this, one possibility is that the APC has mechanisms which facilitate the loading and/or unloading of Ia molecules with processed antigen. We have described one protein whose characteristics suggest it as a candidate to participate in such processes. This is a peptide binding protein which is a member of the heat shock protein (hsp) family. Our evidence for this protein having a role in antigen processing will be reviewed in the following section.

2 Isolation of a Peptide Binding Protein Which Plays a Role in Antigen Processing

The puzzling aspects of the reported binding of peptides to Ia led to a search for peptide binding proteins which might play a role in antigen processing by facilitating the binding of peptide to Ia and/or to the TCR. Detergent-solubilized cell lysates were chromatographed on a peptide affinity column containing a known antigenic peptide of the soluble globular protein antigen pigeon cytochrome c (Pc), residues 81–104 (Pc 81–104). Two to three proteins in the 72–74 kD M_r range are eluted from the column under acid conditions. We refer to these as peptide binding proteins of 72/74 M_r (PBP72/74) (LAKEY et al. 1987). PBP72/74 can be purified from a number of cell types, including B cells, T cells, macrophages, and fibroblasts, and tissues, including spleen and liver, and thus would not appear to be a specialized product of Ia-expressing APCs.

With regard to peptide specificity, PBP72/74 can be competitively eluted from the Pc 81-104 column with the Pc 81-104 peptide but not with intact Pc and thus appears to be specific for some feature of the peptide not found in the native protein. This is the case even though the Pc 81-104 peptide is on the surface of the native protein, exposed to the solvent, and is accessible to binding by specific antibodies. Additional studies indicate that PBP72/74 binding to peptide is not discriminatory of the primary amino acid sequence as other peptides are able to elute PBP72/74 from Pc 81-104, including the corresponding peptide of mouse cytochrome c, and an unrelated peptide of Pc, residues 66-80. In addition, PBP72/74 binds to peptides of the unrelated protein lactate dehydrogenase C4 (LDH-C4). However, the secondary structure of the peptide appears to influence binding to PBP72/74. In preliminary studies it was observed that although a linear peptide of LDH-C4 (residues 315-327) bound PBP72/74, a peptide representing this same region engineered to take on a stable α-helical structure (described in KAUMAYA et al. 1990) did not bind. Thus, there does not appear to be a simple algorithm which predicts peptide binding to PBP72/74 and binding may depend on a characteristic of the peptides free from their native proteins.

To determine whether PBP72/74 plays a role in antigen processing and/or presentation, antibodies were raised in rabbits to the affinity-purified PBP72/74 and tested for their effect on APC function (LAKEY et al. 1987). PBP72/74-specific antisera block, to the 80%–90% level, the processing and presentation of Pc to a Pc-specific T cell hybrid and of ovalbumin (OVA) to an OVA-specific T cell, AODH. The antisera tested are not nonspecifically toxic as these have no effect on the response of the T cell hybrids to the nonspecific mitogen concanavalin A. Blocking appears to be at the level of the APC as T cells incubated with the antisera and washed are able to respond to antigen presented by APCs, while the same treatment of APCs blocks presentation. The blocking activity is not MHC specific in that the activity is absorbed by spleen cells of mice of different MHC haplotypes. Subsequently, monoclonal antibodies which showed the same blocking activity were generated from spleens of rats immunized with the purified PBP72/74. However, such lines have proven to be unusually unstable and failed to secrete antibody after a brief time in culture.

The PBP72/74-specific rabbit antiserum immunoprecipitates PBP72/74 from ^{35}S-labeled cell lysates and from ^{125}I-surface labeled cells and shows a cross-reactivity with monkey proteins of similar molecular weight. By flow cytometry, the PBP72/74-specific antiserum stains the surfaces of B cells and macrophages and does not stain T cells, fibroblasts, or NK cells (VANBUSKIRK et al. 1989; VanBuskirk et al., submitted). The staining of cells does not correlate with the expression of Ia. Indeed, L cells show no cell surface expression of PBP72/74, and L cell lines transfected with the genes encoding the α and β chain of either the I-A or the I-E molecules (Germain et al. 1985; Germain and Quill 1986) (kindly provided by R. Germain, National Institutes of Health) are also negative for PBP72/74 cell surface expression. Moreover, mutations in B cell lymphomas, characterized by Glimcher et al. (1985), which abrogate the cell surface expression of Ia, do not affect the expression of PBP72/74. Taken together, such analyses indicate that PBP72/74 surface expression is a function of B cells and macrophages, independent of Ia expression.

In collaboration with F. Brodsky and co-workers (the University of California, San Francisco) the PBP72/74-specific antisera were used to label PBP72/74 for immunoelectron microscropy in thin sections of the human B cell lymphoma IM-9. The B cells had been incubated with gold-labeled antibodies specific for Ig and allowed to internalize for 2 min. As discussed above, surface Ig is taken into vesicles which contain newly synthesized Ia, and cathepsin B and D, and such vesicles are suggested to represent an intracellular processing site. As compared to control nonspecific immune sera the PBP-specific antisera stained 36% of the Ig-containing vesicles as detected using gold-labeled secondary antibodies. Additional staining of plasma membrane and small cytoplasmic structures was observed. These results indicate that PBP72/74 has a restricted expression within the cell, which is consistent with its playing a role in antigen processing and/or presentation.

3 Identification of PBP72/74 as a Member of the Heat Shock Protein 70 (hsp70) Family

Our initial analysis of PBP72/74 led us to conclude that we had identified a protein which played a role in the processing and/or presentation of antigen by binding to processed antigenic peptides and facilitating their interaction with Ia and/or the TCR. Because PBP72/74 could be purified from a variety of cell types, we considered it unlikely to represent a differentiated product of APCs, although its cell surface expression may be cell type specific. Rather, it was considered that PBP72/74 may belong to a family of proteins, members of which function to bind to denatured or newly synthesized proteins, and subsequently to transport these to an appropriate functional site. One candidate protein was the immunoglobulin binding protein (BiP), which had been described to bind to newly synthesized, unfolded μ heavy chain prior to its folding with the appropriate light chain in the endoplasmic reticulum (Hendershot et al. 1987). However, a monoclonal ahtibody specific for BiP (provided by J. Kearny, University of Alabama, Birmingham) did not recognize purified PBP72/74 in immunoprecipitation although it did precipitate the slightly larger (78 kD Mr) BiP protein from cell lysates. Subsequently, BiP was shown to be a member of the hsp70 family. Using a monoclonal antibody to hsp70 proteins (mAb 7.10) (VELAZQUEZ et al. 1983) which recognizes a number of the hsp70 proteins from a variety of species (provided by S. Lindquist, University of Chicago), PBP72/74 was shown to be serologically related to the hsp70 family. Further serological analysis showed PBP72/74 is recognized in Western blot by the mAb N27 (VASS et al. 1988) (provided by W. Welsh, University of California, San Francisco), specific for a constitutive member of the hsp70 family, but not by mAb N15 (WELCH and SUHAN 1986) (also provided by W. Welsh), which recognizes a stress-induced hsp70 family member. At present, the reactivity of mAb N27 and mAb N15 with B cell or macrophage surfaces has not been tested. However, KAUFMANN and co-workers have reported staining of hsp's on the cell surfaces of stressed host cells (KAUFMANN and WAND-WURTTENBERGER, personal communication, 1990).

PBP72/74 shares another characteristic feature of the hsp70 proteins, which is the ability to bind ATP (MUNRO and PELHAM 1986). The binding of ATP causes the release of the hsp70 proteins from their substrates. Similarly, PBP72/74 binds ATP, which causes the release of PBP72/74 from a peptide column (VANBUSKIRK et al. 1989). The relationship between peptide binding and ATP binding is not known. However, it has been shown that PBP72/74 is eluted from an ATP column by peptide.

At present we do not know whether PBP72/74 is a new member of the hsp70 family.[1] Its subcellular distribution and cell surface expression have not been described for other hsp70 family members. Studies in progress to isolate the gene encoding PBP72/74 in mice should resolve this question. It is not known how PBP72/74 is associated with the cell's plasma membrane. PBP72/74 could have a membrane anchoring domain, which, if so, would be unique among the known hsp70 proteins. Alternatively, PBP72/74 could be membrane associated via a fatty acid tail. However, preliminary studies indicated that lipase treatment of APCs does not alter PBP72/74 cell surface expression. Lastly, PBP72/74 may become membrane associated by binding to another membrane protein. Precedence for this comes from the observation that an hsp70 family member is bound to terminal transferrin receptor in membrane vesicles during reticulo-cyte maturation (DAVIS et al. 1986).

Because PBP72/74 is related to the hsp family, it was of interest to determine whether its expression is regulated by stress or alternatively by activation, which has been shown in lymphocytes to induce transcription of hsp70 (SPECTOR et al. 1989). The cell surface expression of PBP72/74, measured by flow cytometry using the PBP72/74-specific antisera and a fluoresceinated secondary antibody, does not change following incubation of cells at 42 °C for 30 min. PB72/74 cell surface expression is similar in B cells activated with lipopolysaccharide or with the lymphokine IL-4. As stated earlier, KAUFMANN and co-workers report that in mouse macrophages an hsp-specific mAb recognizes a structure which is

---

[1] Amino acid sequence analysis of PBP72 confirms its relationship to the hsp family as it shares the invariant N-terminal region sequence G I D L G and shows that it is indeed an as yet unidentified member of the hsp family.

expressed more strongly on the surfaces of stressed macrophages as compared to untreated macrophages. Although mAb N27 recognizes PBP72/74 in Western blot, we do not know whether it reacts with native surface PBP72/74. It is also not known whether the regulation of PBP72/74 is similar in B cells and in macrophages. WINFIELD and co-workers have also described the expression of hsp70 proteins on activated human HL60 cells (JARJOUR et al. 1989). In this case surface expression was detected using human antibodies to hsp70 proteins found in the serum of autoimmune individuals. We observed that the same human antisera recognized PBP72/74 in Western blot but have not tested them for staining of mouse B cell surfaces (unpublished observation).

4 Comments on a Role for Heat Shock Proteins in Antigen Processing and Presentation and in Autoimmunity

Our results indicate that a heat shock protein expressed on the cell surface and in endosomes is linked to the processing and presentation of antigen by the observation that antibodies specific for it block APC function. Current knowledge of the function of the heat shock proteins indicates that they are chaperones binding to newly synthesized polypeptides before folding occurs so as to prevent inappropriate interactions among chains. In some cases heat shock proteins serve to transport these to a functional site in the cell. In this context, one could envision that members of this family function in binding newly produced peptides in the processing compartments. The function of heat shock proteins, such as PBP72/74, may be to scavenge peptides from degradative compartments, thereby preventing complete proteolysis, and concentrating these for binding to Ia. It is possible that PBP72/74 binds the released peptides and transports these, from the degradative compartment, to a compartment where Ia binds peptide, if these are not one and the same. A portion of PBP72/74 associated with peptide may be transported to the cell surface in Ia-containing vesicles to display peptides for Ia or TCR binding. In each of these cases hsp70 proteins would play a scavenging and concentrating role in antigen processing. It is also possible that the hsp70 proteins play a more fundamental role in facilitating Ia–peptide binding. In this regard, the observation by CRESSWELL (1990) that Ia bound to its invariant chain is not accessible for peptide binding is of interest. The invariant chain is proteolytically cleaved from the Ia molecule at a late stage in synthesis prior to cell surface expression. During such a process surfaces of the Ia molecule which were bound to invariant chain must be exposed to the environment. Heat shock proteins may function to bind to the exposed sites on Ia, allowing for appropriate folding processes which could include peptide binding. Such a function for heat shock proteins could be in addition to the simpler scavenger/concentration function. It is intriguing that the hsp70 gene is located in the MHC complex in man and thus is genetically linked to the MHC proteins (SARGENT et al. 1989). Future studies may clarify whether there is a functional link between the MHC and the hsp70 proteins.

Several lines of evidence indicate a role for heat shock proteins in autoimmunity, as reviewed in this volume and in KAUFMANN (1990) and YOUNG (1990). The general theme of such observations is that T cells specific for heat shock proteins of pathogenic organisms cross-react by an as yet undefined mechanism with self-hsp proteins, causing the autoimmune phenomenon. Whether PBP72/74 is a target of hsp-specific T cells is not known. If hsp-specific T cells recognize antigen in a conventional MHC-restricted fashion then cell surface expression of native PBP72/74 is inconsequential. For subpopulations of T cells, such as $\gamma\delta$ T cells, which may not be MHC restricted in their recognition of antigen, PBP72/74 may represent a target. At present, while the observations implicating heat shock proteins in autoimmunity are compelling, too many pieces of the puzzle are missing to understand how these all might fit together.

References

Amit AG, Mariuzza RA, Phillips SEV, Poljak RJ (1986) Three-dimensional structure of an antigen–antibody complex at 2.8 A resolution. Science 233: 747–753

Babbitt BP, Allen PM, Matsueda G, Haber E, Unanue ER (1985) Binding of immunogenic peptides to Ia histocompatibility molecules. Nature 317: 359–360

Bjorkman PJ, Saper MA, Samraoui B, Bennett WS, Strominger JL, Wiley DC (1987) Structure of the human class I histocompatibility antigen, HLA-A2. Nature 329: 506–512

Buus S, Sette A, Colon SM, Jenis DM, Grey HM (1986) Isolation and characterization of antigen–Ia complexes involved in T cell recognition. Cell 47: 1071–1077

Cresswell P (1990) Questions of presentation. Nature 343: 593–594

Davis JQ, Dansereau D, Johnstone RM, Bennett V (1986) Selective externalization of an ATP-binding protein structurally related to the clathrin-uncoating ATPase/heat shock protein in vesicles containing terminal transferrin receptors during reticulocyte maturation. J Biol Chem 261: 15368–15371

Davis MM, Bjorkman PJ (1988) T-cell antigen receptor genes and T-cell recognition. Nature 334: 395–402

Germain R, Quill H (1986) Unexpected expression of a unique mixed type class II MHC molecule by transfected L cells. Nature 320: 72–75

Germain RN, Ashwell JD, Lechler RA, Margulies DH, Nickerson KM, Suzuki G, Tou JYL (1985) "Exon-shuffling" maps control of antibody and T cell recognition sites to the NH2 terminal domain of the class II major histocompatibility polypeptide A_β. Proc Natl Acad Sci (USA) 82: 2940–2944

Glimcher LH, McKean DJ, Choi E, Seidman JG (1985) Complex regulation of class II gene expression analysis with class II mutant cell lines. J Immunol 135: 3542–3550

Guagliardi LE, Koppelman B, Blum JS, Marks MS, Cresswell P, Brodsky FM (1990) Co-localization of molecules involved in antigen processing and presentation in an early endocytic compartment. Nature 343: 133–139

Harding CV, Leyva-Cobian F, Unanue ER (1988) Mechanisms of antigen processing. Immunol Rev 106: 77–92

Hendershot L, Bole D, Kohler G, Kearney JF (1987) Assembly and secretion of heavy chains that do not associate posttranslationally with immunoglobulin heavy chain-binding protein. J Cell Biol 104: 761–767

Jacobson S, Sekaly RP, Jacobson CL, McFarland HF, Long EO (1989) HLA class II-restricted presentation of cytoplasmic measles virus antigens to cytotoxic T cells. J Virol 63: 1756–1762

Jarjour W, Tsai V, Woods V, Welch W, Pierce S, Shaw M, Mehta H, Dillmann W, Zvaifler N, Winfield J (1989) Cell surface expression of heat shock proteins. Arthritis Rheum 32: S44

Kaufmann SHE (1990) Heat shock proteins and the immune response. Immunol Today 11: 129–136

Kaumaya PTP, Berndt KD, Heidorn DB, Trewhella J, Kezdy FJ, Goldberg E (1990) Synthesis and biophysical characterization of engineered topographic immunogenic determinants with αα topology. Biochemistry 29: 13–23

Lakey EK, Margoliash E, Pierce SK (1987) Identification of a peptide binding protein that plays a role in antigen presentation. Proc Natl Acad Sci 84: 1659–1663

Lakey EK, Casten LA, Niebling WL, Margoliash E, Pierce SK (1988) Time dependence of B cell processing and presentation of peptide and native protein antigens. J Immunol 140: 3309–3314

Lanzavecchia A (1990) Receptor-mediated antigen uptake and its effect on antigen presentation to class II restricted T lymphocytes. Ann Rev Immunol 8: 773–793

Lee JM, Watts TH (1990) On the dissociation and reassociation of MHC class II-foreign peptide complexes. J Immunol 144: 1829–1834

Malissen B, Peele P, Goverman JM, McMillan M, White S, Kappler J, Marrack P, Pierres A, Pierres M, Hood G (1984) Gene transfer of H-2 class II genes: antigen presentation by mouse fibroblast and hamster B-cell lines. Cell 36: 319–327

Matis LA (1990) The molecular basis of T cell specificity. Ann Rev Immunol 8: 65–82

Morrison LA, Lukacher AE, Braciale VL, Fan DP, Braciale TJ (1986) Differences in antigen presentation to MHC class I- and class II-restricted influenza virus-specific cytolytic T lymphocyte clones. J Exp Med 163: 903–921

Munro S, Pelham HRB (1986) An hsp70-like protein in the ER: identity with the 78 Kd glucose-regulated protein and immunoglobulin heavy chain binding protein. Cell 46: 291–300

Neefjes JJ, Stollorz V, Peters PJ, Geuze HJ, Ploegh HL (1990) The biosynthetic pathway of MHC class II but not class I molecules intersects the endocytic route. Cell 61: 171–183

Roosnek E, Demotz S, Corradin G, Lanzavecchia A (1988) Kinetics of MHC-antigen complex formation on antigen-presenting cells. J Immunol 140: 4079–4082

Sadegh-Nasseri S, McConnell HM (1989) A kinetic intermediate in the reaction of an antigenic peptide and I-E^K. Nature 337: 274–276

Sargent CA, Dunham I, Trowsdale J, Campbell RD (1989) Human major histocompatibility complex contains genes for the major heat shock protein HSP70. Proc Natl Acad Sci 86: 1968–1972

Spector NL, Freedman AS, Freeman G, Segil J, Whitman JF, Welch WJ, Nadler LM (1989) Activation primes human B lymphocytes to respond to heat shock. J Exp Med 170: 1763–1768

Srinivasan M, Pierce SK (1990) Isolation of a functional antigen-Ia complex. Proc Natl Acad Sci USA 87: 919–922

VanBuskirk A, Crump BL, Margoliash E, Pierce SK (1989) A peptide binding protein having a role in antigen presentation is a member of the hsp70 heat shock family. J Exp Med 170: 1799–1809

Vass K, Welch WJ, Nowak TS (1988) Localization of 70 kDa stress protein in gerbil brain after ischemia. Acta Neuropathol 77: 128–132

Velazquez JM, Sonoda S, Bugaisky G, Lindquist S (1983) Is the major drosophila heat shock protein present on cells that have not been heat shocked? J Cell Biol 96: 286–290

Weiss S, Bogen B (1989) B-lymphoma cell process and present their endogenous immunoglobulin to major histocompatibility complex-restricted T cells. Proc Natl Acad Sci USA 86: 282–286

Welch WJ, Suhan JP (1986) Cellular and biochemical events in mammalian cells during and after recovery from physiological stress. J Cell Biol 103: 2035–2052

Young RA (1990) Stress proteins and immunology. Ann Rev Immunol 8: 401–420

Heat Shock Proteins and Inflammation

B. S. POLLA and S. KANTENGWA

1 Introduction

Inflammation is produced by the accumulation of circulating cells in a particular tissue or organ. The "*actors*" in inflammation thus are the inflammatory cells, neutrophils, eosinophils, monocytes–macrophages, platelets, and their products, lipid mediators, cytokines, and oxygen free radicals; the *targets* of the inflammatory process are the cells of the inflamed tissue, as well as the inflammatory cells themselves, and the extracellular matrix. n this chapter, we will review how heat shock proteins (hsp's) have been and are linked to inflammation—for example through fever. We will also discuss the relationships between hsp's and *the actors in inflammation* as well as their interaction with a target, the extracellular matrix. Finally, we will allude to the possible role(s) of hsp's in inflammatory diseases and anti-inflammatory therapies.

For a long time in medical history, heat has been linked to inflammation in two ways. Heat, "calor," was recognized as a cardinal sign of inflammation by

Allergy Unit, University Hospital, 1211 Geneva 4, Switzerland

Current Topics in Microbiology and Immunology, Vol. 167
© Springer-Verlag Berlin · Heidelberg 1991

Celciùs: "*calor*, dolor, rubor, tumor." In addition heat was proposed as a possible therapeutic approach in inflammatory and other diseases; thus Hippocrates said: "Give me the possibility to produce fever and I will heal the diseases...." We would like to suggest that fever is a natural way to generate heat, and therefore to induce in vivo synthesis of hsp's. Fever could act as a stimulator of the immune response when necessary, but also, *via* hsp synthesis, represent a means of physiological feedback control of the inflammatory process.

2 Fever and Heat Shock Proteins

KLUGER has performed a great deal of work to substantiate the value of fever as proposed by Hippocrates (KLUGER et al. 1975; KLUGER 1986). When the desert lizard is infected with *Aeromonas hydrophila*, the animal seeks the warmest environment in the sand so as to increase its body temperature to 40–41 °C (goldfish display similar behavior in water). When this natural behavior—and thus the rise in body temperature—is prevented, the animals die from their infection, and the difference in survival between the two groups ("febrile" vs "non-febrile") is highly significant (KLUGER 1986). One should also recall that prior to the antibiotic era, fever therapy and local heating were advocated for the treatment of infectious diseases such as syphilis and gonorrhea (for review, see MALKINSON 1980).

Whether this protective effect of fever is related to in vivo synthesis of hsp's has not yet been investigated. Elevation in body temperature has been shown to induce in vivo hsp synthesis in several animals, regardless of whether the elevation in body temperature was obtained by physical or pharmacological means. In humans, the effects of fever on hsp synthesis have not previously been examined, and it is thanks to Corinne Reynaud, MD, a devoted intern who remained at work despite a flu-like illness and a high fever (39.3 °C for 2 h), that we have been able to demonstrate that fever does modulate the heat shock response in circulating neutrophils. As shown in Fig. 1, not only was there in vivo induction of a low grade hsp synthesis (compare lane 6 and lane 5 for hsp70 synthesis), but also a differential protein synthesis when cells were further exposed to elevated temperatures in vitro (compare lanes 4 and 14). When neutrophils from normal donors were exposed in vitro to 45 °C for 20 min there was a complete cessation of normal protein synthesis (Fig. 1, lane 4, and MARIDONNEAU-PARINI et al. 1988). In contrast, synthesis of hsp's was increased and normal protein synthesis partially maintained in cells previously exposed in vivo to 39.3 °C (Fig. 1, lane 14). These cells thus appear to have acquired in vivo thermotolerance. This experiment may represent the first evidence that hsp's are indeed part of the fever response in humans, and that fever may exert its beneficial effects at least in part *via* modulation of the heat shock response.

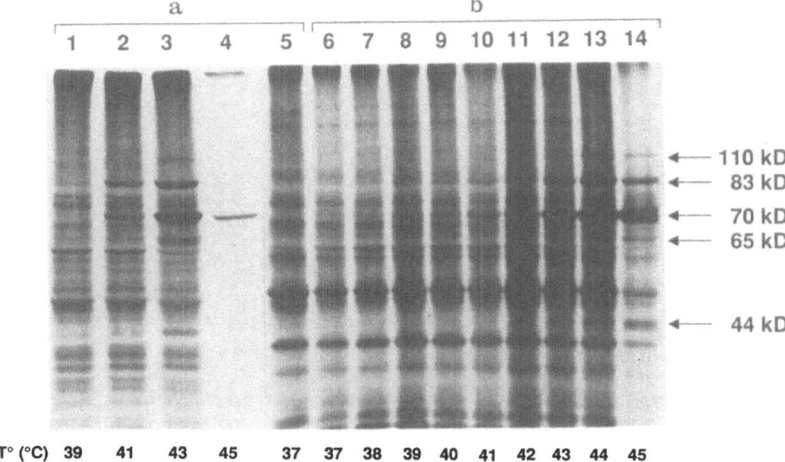

Fig. 1. Autoradiograph of a polyacrylamide gel electrophoresis of proteins synthesized by polymorphonuclear leukocytes from normal controls (**a**) and a patient with fever (39.3°C) associated with a common flu-like syndrome and no other disease (**b**). Cells were prepared, further exposed for 20 min to the indicated temperatures in vitro, and labeled with ^{35}S-methionine as previously described (MARIDONNEAU-PARINI et al. 1988). The induction of hsp70 (*lane 6*) and the maintenance of normal protein synthesis (*lane 14*) indicate modulation of the heat shock response by fever— although we have not controlled for the effects of a flu-like syndrome *without* fever

3 Lipid Mediators of Inflammation and Heat Shock Proteins

The major lipid mediators of inflammation are shown in Fig. 2. Their role in inflammation is underlined by the design of available anti-inflammatory therapies: corticosteroids probably exert their effects *via* inhibition of phospholipase A$_2$, and most clinically relevant nonsteroidal anti-inflammatory drugs (NSAIDs) are inhibitors of cyclooxygenase; PAF-acether antagonists are still in the experimental phase. Prostaglandins, and in particular prostaglandin E$_2$ (PGE$_2$), are produced by activated neutrophils and monocytes–macrophages and play a role in the tissue destruction and repair associated with chronic inflammation such as arthritis. Leukotrienes and PAF-acether are predominantly involved in acute inflammation and allergy. HIGHTOWER was the first to specifically relate hsp's to inflammation (HIGHTOWER and WHITE 1981). He suggested that phenomena associated with inflammatory responses may provide a fertile research ground for the elusive functions of hsp's, and proposed an association between the metabolites of arachidonic acid and the stress response. We have shown that heat shock inhibits NADPH oxidase activity in human neutrophils, whereas the generation of PGE$_2$ is unaffected

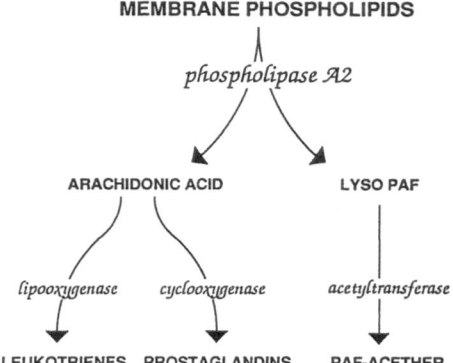

MEMBRANE PHOSPHOLIPIDS

phospholipase A2

ARACHIDONIC ACID   LYSO PAF

lipooxygenase   *cyclooxygenase*   *acetyltransferase*

LEUKOTRIENES PROSTAGLANDINS   PAF-ACETHER

Fig. 2. Leukotrienes, prostaglandins, and PAF-acether are derived from membrane phospholipids by distinct enzymatic pathways. The effects of heat shock on prostaglandin and leukotriene production have been investigated by us and by others (see text), and Fig. 3b shows the effects of PAF-acether on protein synthesis. The relevance of these lipid mediators in inflammation is underlined by the fact that most anti-inflammatory drugs (either steroids or NSAIDs—as well as heat shock!) inhibit enzyme(s) involved in this metabolism

(MARIDONNEAU-PARINI et al. 1988). The inhibitory effects of heat shock on the generation of oxygen metabolites has been confirmed by KOLLER et al. (1989) using chemoluminescence rather than the superoxide dismutase (SOD) inhibitible reduction of ferricytochrome c (MARIDONNEAU-PARINI et al. 1988). These authors also investigated the effects of heat shock on the lipooxygenase pathway, and found that preexposure to heat shock (42 °C for 60 min) inhibited the calcium ionophore-stimulated generation of leukotrienes. In contrast, heat shock resulted by itself, without further stimulation, in the conversion of ^{14}C-arachidonic acid into LTB_4, 5-HPETE, and 5-HETE (KOLLER et al. 1989). Rather than the effects of heat shock on the metabolism of arachidonic acid into PAF-acether, we investigated the effects of PAF-acether on protein synthesis by human monocytes. PAF-acether activates the respiratory burst in these cells, and we have previously reported that NADPH oxidase activation during phagocytosis of red blood cells induces hsp synthesis in monocytes–macrophages (CLERGET and POLLA 1990, and Fig. 3a; see below). As shown in Fig. 3b, PAF-acether did not induce HSP synthesis, even when tested over a large range of concentrations and incubation times. Above 10^{-4} M, PAF-acether was toxic to the cells and protein synthesis was abolished, but without induction of the classical hsp's (see for comparison Fig. 3a). PAF-acether did, however, induce the synthesis of a protein of apparent molecular weight 78 kD, a protein which we also found to be induced by exposure of monocytes–macrophages to opsonized zymosan (Fig. 3a, lane 5). Whether this latter protein is or is not part of the stress protein family (grp78?) remains to be determined. In any case, the reported beneficial effects of heat in inflammatory conditions such as arthritis could be related to partial inhibition of the

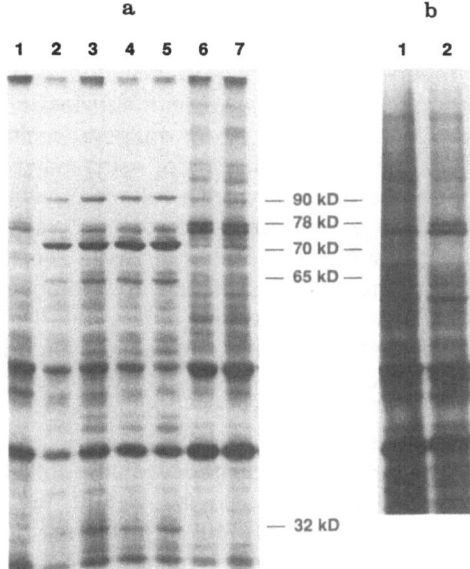

a

| 1 | 2 | 3 | 4 | 5 | 6 | 7 |

b

| 1 | 2 |

— 90 kD —
— 78 kD —
— 70 kD —
— 65 kD —

— 32 kD

Fig. 3. Effects of phagocytosis, activation with opsonized zymosan, and iron (A) or PAF-acether (B) on the heat shock response in human monocytes. Peripheral blood monocytes from normal donors were isolated by gradient centrifugation and purified by adherence, then exposed (a) to oSRBCs as described (CLERGET and POLLA 1990) (*lanes 2-5*) in the presence (*lanes 3-5*) cr absence (*lane 2*) of the iron chelator desferrioxamine (10, 20, and 50mM respectively), or to opsonized zymosan (300 mg/ml) with (*lane 6*) or without (*lane 5*) the addion of exogenous iron (100 µM), or (**b**) to PAF-acether (*lane 2*, 10^{-4} M). Incubations were for 3 h and labeling for 90 min. *Lanes 1*, **a** and **b**: controls. The classical hsp's (70, 90, 65 kD) and heme oxygenase (32 kD) were induced during erythrophagocytosis, and their induction was not prevented by desferrioxamine; in contrast, exposure to osponized zymosan resulted in the induction of a protein of apparent molecular weight 78 kD, which was also induced by PAF-acether. The addition of exogenous iron to opsonized zymosan resulted in slight induction of hsp70, further suggesting a role for iron in induction of hsp's during phagocytosis, related to the catalysis of hydroxyl radical formation

arachidonic acid cascade. The relationships between this metabolism and hsp's deserve further investigation.

4 Cytokines, Inflammation, and Heat Shock Proteins

4.1 Interleukin-1 and Tumor Necrosis Factor α

In 1983, DUFF and DURUM reported that the mitogenic and pyrogenic effects of interleukin-1 (IL-1) were linked. Indeed, IL-1-induced lymphocyte proliferation was increased when lymphocytes were incubated with IL-1 at elevated temperatures, and this effect appeared to be specific for IL-1, since lipopolysaccharide-mediated proliferation was unaffected under similar culture conditions. We and others are investigating the intriguing possibility that hsp's were involved in this observation, by inducing hsp synthesis in lymphocytes or in the antigen-presenting cells in either MLR cultures or in antigen-specific lymphocyte proliferation assays. Indeed, one of the functions of hsp's may be antigen processing and presentation; another chapter in this volume deals in detail with the issue. Heat shock may also increase the expression of MHC class II molecules (A. REES et al., our unpublished observations). Returning to IL-1, there is a single report suggesting that IL-1 does induce hsp synthesis in isolated rat islets of Langerhans (HELQVIST et al. 1989) whereas we and others found no induction of hsp's by IL-1 in any other cell examined. To the best of our knowledge there is to date no report that tumor necrosis factor α (TNF) induces the synthesis of hsp's, whereas both IL-1 and TNF induce phosphorylation—but not synthesis!—of the low molecular weight hsp's (KAUR et al. 1989). Because no clear signal transduction pathway has yet been identified in particular for IL-1, the possibility that phosphorylation of the low molecular weight hsp's participates in signalling by IL-1 has been considered. TNF induces another protective protein in respect of which there is still discussion as to whether it is part of the hsp family: the scavenging enzyme SOD is induced by TNF and appears essential in the protective effects against TNF-mediated cell lysis (WONG et al. 1989). In bacteria, SOD is a recognized hsp since it has been shown to be induced by heat shock (PRIVALLE and FRIDOVICH 1987); we are currently reevaluating this issue in human cells. Finally, it has recently been reported that heat shock protects cells from TNF-induced lysis (JÄÄTTELÄ et al. 1989).

4.2 Interleukin-2

Not only does interleukin-2 (IL-2) induce accumulation of hsp-70-specific mRNA and the synthesis of hsp70 in IL-2 receptor-bearing cells (FERRIS et al. 1988), but the IL-1-driven secretion of IL-2 has also been shown to be highly temperature dependent (LEDERMAN et al. 1987). GRANELLI-PIPERNO et al. (1986) reported previously that various cellular activators [including phytohemagglutinin A and phorbol myristate acetate (PMA)] induced accumulation of hsp70-specific mRNA in human T cells, along with c-*myc*. Several reports have indicated that hsp expression is linked to expression of oncogenes important in cellular activation, such as *fos* (MÜLLER 1986; ANDREWS et al. 1987), but the precise relationship between cellular activation, oncogene expression, and hsp induction remains to be determined (KINGSTON et al. 1985). In the particular case of *fos*, heat shock induces an accumulation of *fos* RNA; this accumulation, however, is not related to increased transcription of the oncogene, but rather to the general inhibition of normal cellular protein synthesis, including a short-lived repressor of *fos* RNA expression (ANDREWS et al. 1987), although heat shock consensus sequences (GAA-TTC) are present in the regulatory sequences of *fos* (see MÜLLER 1986).

4.3 Interferons

Interferon is another of the endogenous pyrogens, as are IL-1, IL-2, and TNF. Interferons-α and -β modulate the heat shock response in murine cells; in particular, they lower the threshold for maximal induction of hsp's by heat (TUMARKIN et al. 1985; MORANGE et al. 1986). On the other hand, elevated temperatures induce interferon-γ activity in human B lymphoblastoid lines (TAYLOR et al. 1984). More recently, KAUFMANN's group reported that treatment of murine bone marrow derived macrophages with interferon-γ stresses the cells in such a way that they become targets for T cells specific against a bacterial heat shock protein (KOGA et al. 1989). We, however, found no effects of pretreatment of the human premonocytic line U937 with interferon-γ for 72 h on the subsequent effects of heat shock, whereas differentiation of these cells with another agent, the steroid hormone 1,25-dihydroxyvitamin D_3 [1,25-$(OH)_2D_3$], markedly increases the synthesis of hsp's and protects the cells from thermal injury (see below).

4.4 Platelet-Derived Growth Factor

Platelet-derived growth factor (PDGF) is an important competence factor in cell proliferation and activation, and has been shown to be involved in the pathogenesis of inflammatory lesions, in particular in the lung. In his early paper in which he referred to inflammation as "fertile ground" for investigation of the functions of hsp's, HIGHTOWER mentioned induction of hsp's by PGDF (partially purified human PDGF, in chicken and rat embryo cells) (HIGHTOWER and WHITE 1981). We have reinvestigated this issue using human monocytes as target cells and recombinant human PDGF (Sigma) but found no induction of hsp's at any concentration of PDGF tested (from 1.5 to 15 ng/ml) or with either incubation time (3 or 24 h).

5 Oxygen Free Radicals, Phagocytosis, and Heat Shock Proteins

Whereas oxygen free radicals have long been regarded as classical inducers of hsp's in bacteria, it is only recently that they have been recognized as such in human cells (for review see DONATI et al., in press). Reactive oxygen species have the ability to alter proteins and induce DNA strand breaks, and the presence of abnormal proteins is the most likely signal for hsp induction (ANANTHAN et al. 1986).

We have been particularly interested in the role of oxygen free radicals as inducers of hsp's in *phagocytic cells*, since these cells are equipped with an enzymatic system for the generation of such reactive oxygen species, i.e., the respiratory burst enzyme NADPH oxidase. This enzyme, which generates superoxide anion (O_2^-) from molecular oxygen at the expense of NADPH, can be activated through distinct pathways by various soluble or particulate stimuli (MARIDONNEAU-PARINI et al. 1986). NADPH oxidase is activated during phagocytosis, and in this situation the toxic oxygen metabolites contribute to the host defenses by participating in bacterial killing. Oxygen free radicals, however, can also be toxic to the host and their role in many different types of inflammatory processes and human diseases is widely accepted (HALLIWELL 1987). The potential preventive and/or curative effects of antioxidants have been investigated in an array of different inflammatory diseases such as rheumatoid arthritis and the adult respiratory distress syndrome. The limited success of antioxidants as therapeutic tools may be related to pharmacological problems such as bioavailability rather than to the rationale behind their use.

Increasing lines of evidence suggest that hsp's could act as endogenous antioxidants, and manipulation of this system may open new avenues to antioxidant therapies (POLLA 1988; DONATI et al. 1990). The possibility that hsp's could offer protection against oxidative injury has been suggested in bacteria by CHRISTMAN et al. (1985), who reported that mutant *Salmonella* resistant to oxidative stress spontaneously overexpress an array of hsp's. In phagocytes, hsp's and the oxidation-specific stress protein, p32, or heme oxygenase (KEYSE and TYRRELL 1989)—induced by phagocytosis of opsonized sheep red blood cells (oSRBC)s—may be part of the phagocytes' protective mechanisms against the toxic products they themselves produce (CLERGET and POLLA 1990).

The mechanisms by which hsp's are induced during phagocytosis appear complex and may require the presence of *both* the "nonself" (phagocytosed) material and the toxic oxygen metabolites. To examine further the roles of oxygen free radicals and phagocytosis in hsp synthesis, we compared protein synthesis in monocytes from normal donors (Fig. 4A) and patients with chronic granulomatous disease (Fig. 4B). Indeed, because of genetic alteration or lack of one or the other component of the NADPH oxidase, these patients are unable to produce O_2^-, whether their neutrophils or monocytes are activated by PMA or during phagocytosis (EZEKOWITZ et al. 1987). Phagocytosis is conserved; before the new forms of the therapy, including interferon-γ, became available, these patients nevertheless died of pyogenic infections in early childhood (EZEKOWITZ et al. 1987). As shown in Fig. 4, synthesis of hsp's is conserved in the cells from these patients, whether the inducer is heat shock (lanes 6) or phagocytosis of oSRBCs (lanes 2, 3, 4). Since oxygen free radicals (and in particular hydroxyl radicals, see also legend to Fig. 3) appear from our earlier work (CLERGET and POLLA 1990) to be a requirement for hsp induction during erythrophagocytosis, these results suggest that hydroxyl radicals are generated in monocytes from patients with chronic granulomatous disease by pathways other than NADPH oxidase-mediated O_2^- production. It also appears from this experiment that the phorbol ester PMA, a potent inducer of the respiratory burst in neutrophils and monocytes—macrophages, does not by itself induce hsp synthesis in these cells, in contrast to its reported effects in lymphocytes (see above).

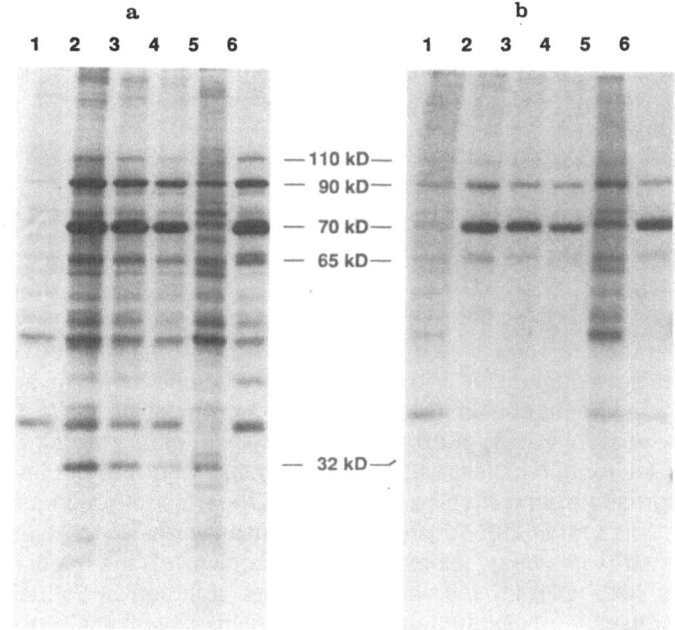

Fig. 4. Normal human monocytes (**a**) and monocytes from a patient with chronic granulomatous disease (CGD) (**b**) (2×10^6 cells/condition) were exposed to 45°C for 20 min (*lanes 6*), to PMA (100 ng/ml) (*lanes 5*), or to oSRBCs (*lanes 2*, 200 μl; *lanes 3*, 100 μl) or non-opsonized SRBCs (200 μl) (*lanes 4*) for 3 h, then labeled for 90 min with ^{35}S-methionine. The classical hsp's (70, 90, 110, 65 kD) and heme oxygenase (32 kD) were induced by erythrophagocytosis in both normal and CGD monocytes; heat shock induces in human cells the classical hsp's but not heme oxygenase. PMA had no major qualitative effect on protein synthesis in either set of cells

6 Metallothioneins and Inflammation

Metallothioneins are part of the stress protein family, as are hsp's, grp's, ubiquitin, and other proteins with antioxidant properties such as heme oxygenase and possibly SOD. Metallothioneins are cysteine-rich, low molecular weight proteins that selectively bind heavy metal ions, such as cadmium and zinc; it has been suggested that metallothioneins protect against heavy metal toxicity and act as free radical scavengers. They are induced by a variety of stresses, including heat, but also IL-1, bacterial endotoxins, and 1,25-(OH)$_2$D$_3$ (DURNAM et al. 1984; KARASAWA et al. 1987). Metallothioneins thus appear to be even more responsive than *hsp* genes to some important mediators of inflammation (IL-1, bacterial endotoxins); since they respond to the latter, the possibility that—in contrast to the classical hsp's but as is the case for SOD—they are induced by TNF as well should be considered. Metallothioneins may belong to the "acute phase proteins"; however, their general role in inflammation and stress remains even more elusive than that of the classical hsp's.

7 The Extracellular Matrix, Inflammation, and Heat Shock Proteins

Adhesion to the extracellular matrix is an essential function of cells both in normal growth and development and in inflammation, wound healing, and tumor cell invasion. Inflammation is characterized by connective tissue remodelling, and deposition of excess or abnormal extracellular matrix proteins is part of this process. Interaction of cells with their extracellular matrix modulates their attachment, spreading, and motility. Interestingly, heat shock *decreases* cellular adhesion (POLLA et al. 1986), probably because of the cytoskeletal alterations associated with heat shock (i.e., breakdown of intermediate filaments), whereas 1,25-$(OH)_2D_3$ increases monocyte adherence, induces collagen binding proteins, and simultaneously *increases* hsp synthesis (POLLA et al. 1986; POLLA et al. 1987a, b). On the other hand, a major 47-kD chick embryo fibroblast collagen binding protein *is* an hsp (NAGATA et al. 1986); this cell type-specific, heat shock, and transformation-sensitive protein has been localized to the ER, where it could modulate retention or secretion of extracellular matrix proteins or be involved in intracellular degradation of procollagen. Another interesting link between cellular adhesion and the heat shock system is the homology between the high endothelial venules' lymphocyte receptor and ubiquitin (ST. JOHN et al. 1986). Mature lymphocytes circulate throughout the body and pass through the lymphatic organs such as peripheral lymph nodes, where they specifically recognize and adhere to the high endothelial cells of the postcapillary venules (ST. JOHN et al. 1986). Ubiquitin is part of the receptor by which lymphocytes home in and enter the lymph nodes.

Atherosclerosis is one example of a physiopathological process leading to altered extracellular matrix, in which are involved the cells and mediators of inflammation: monocytes–macrophages, platelets, soluble products of these cells such as oxygen free radicals, lipid mediators (e.g., thromboxanes, PAF-acether), and cytokines such as PDGF, all of them contributing in some way to the development of the atherosclerotic plaque. Increased expression of hsp70 is found in atherosclerotic plaques, where there is intense remodelling of extracellular matrix as well as the presence of platelets, clots, and macrophages (BERBERIAN et al. 1990); hsp70 is concentrated around sites of necrosis and in association with lipid-laden macrophages. The overexpression of hsp70 in atherosclerotic plaques is probably secondary to locally produced oxygen free radicals, rather than PDGF or other macrophage-derived cytokines.

8 Inflammatory Diseases, Anti-inflammatory Drugs, and Heat Shock Proteins

Arthritis may be considered as an inflammatory and/or an autoimmune disease. Hsp's in arthritis as an autoimmune disease are discussed elsewhere in this volume. One of the first papers to have described hsp's in arthritis was from the inflammatory point of view: KUBO et al. reported in 1985 that chondrocytes from patients with osteoarthritis spontaneously synthesized the 70-kD and 90-kD hsp's in vitro, suggesting in vivo activation of the heat shock genes by the inflammatory process. Moreover, the degree of hsp synthesis appeared related to the degree of activity of the osteoarthritis. In this context, the synthesis of hsp's by chondrocytes may be viewed as a consequence of local heat [*calor* still remains a major clinical sign of (osteo) arthritis] or of the local generation, in the joints, of mediators of inflammation, in particular oxygen free radicals.

In support of the inflammatory hypothesis of involvement of hsp's in human disease is the recent observation that hsp's and heme oxygenase are spontaneously, ex vivo, synthesized by alveolar macrophages in certain types of inflammatory lung diseases associated with a high percentage of eosinophils in the bronchoalveolar lavage (POLLA et al. 1989). Eosinophils are, among the polymorphonuclear leukocytes, toxic not only via generation of oxygen free radicals, which they share with the other phagocytes, but also *via* specific eosinophilic proteins. Exposure of monocytes–macrophages to purified proteins from the eosinophil (major basic protein eosinophil cationic protein, eosinophil-derived neurotoxin) will allow the determination of which product(s) of this highly toxic cell is (are) indeed involved in hsp synthesis by the alveolar macrophage. The role hsp's induced in other cells by the presence of eosinophils may play in inflammatory diseases classically associated with the presence of this cell, such as *asthma*, represents an exciting new approach to another old disease.

Another link between inflammatory diseases and hsp's is actually represented by some of the classical anti-inflammatory therapies—and not only heat! Steroids are among the classical modulators of the heat shock response in *Drosophila* and induce metallothioneins. In human cells, however, we found no effect of dexamethasone on the heat shock response (POLLA et al. 1987c) whereas $1,25-(OH)_2D_3$ increases hsp synthesis upon exposure to heat and modulates thermotolerance in these same cells. In this respect one should recall that the 90-kD hsp, which can associate with tyrosine kinases, is also a constituent of the steroid hormone receptor complex. However, the potential implications of this observation for the role of steroids in the regulation of the heat shock response remain entirely speculative.

Probably of more relevance to anti-inflammatory therapy is the induction of heme oxygenase by gold compounds used in the treatment of rheumatoid arthritis (CALTABIANO et al. 1986). The possibility that stress proteins, and in particular heme oxygenase, exert some protective or therapeutic effect *via* their antioxidant potential (and not only *via* induction of immunological tolerance) is one of our favored hypotheses—but still a hypothesis.

Our conclusions may concur with those of numerous others. However, many more data are needed to unravel the precise role(s) of heat and heat shock proteins in inflammation. If this review somehow stimulates the generation of such data, our goal will have been attained.

Acknowledgments. We are grateful to P.D. Lew, MD, for providing the cells from patients with chronic granulomatous disease, to M. Clerget, PhD, for his contribution to the early aspects of this work, to F. Villars for technical assistance, and to G. Rapin and P. Sardy for the artwork. Experimental work mentioned and illustrated here has been supported by the Swiss National Research Foundation (grant 3.960-0.87 to B.S.P.) and by the Foundation pour Recherches medicales Carlos et Elsie de Reuter. S.K. was supported by the Sir Jules Thorn Charitable Trust.

References

Ananthan J, Goldberg AL, Voellmy R (1986) Abnormal proteins serve as eukaryotic stress signals and trigger the activation of heat shock genes. Science 232: 522–524

Andrews GK, Harding MA, Calvet JP, Adamson ED (1987) The heat shock response in HeLa cells is accompanied by elevated expression of c-*fos* proto-oncogene. Mol Cell Biol 7: 3452–3458

Berberian PA, Myers W, Tytell M, Challa V, Bond MG (1990) Immunohistochemical localization of heat shock protein-870 in normal-appearing and atherosclerotic specimens of human arteries. Am J Pathol 136: 71–80

Caltabiano M, Koestler TP, Poste G, Greig RG (1986) Induction of mammalian stress proteins by triethlyphosphine gold compound used in the therapy of rheumatoid arthritis. Biochem Biophys Res .Comm 138: 1074–1080

Christman MF, Morgan RW, Jacobson FS, Ames BN (1985) Positive control of a regulon for defenses against oxidative stress and some heat shock proteins in *Salmonella typhimurium*. Cell 41: 753–762

Clerget M, Polla BS (1990) Erythrophagocytosis induces heat shock protein synthesis by human monocytes-macrophages. Proc Natl Acad Sci USA 87: 1081–1085

Donati YRA, Slosman DO, Polla BS (1990) Oxidative injury and the heat shock response. Biochem Pharmacol (in press)

Duff GW, Durum SK (1983) The pyrogenic and mitogenic actions of interleukin-1 are related. Nature 304: 449–451

Durnam DM, Hoffman JS, Quaife CJ, Benditt EO, Chen HY, Brinster RL, Palmiter RD (1984) Induction of mouse metallothionein-I mRNA by bacterial endotoxin is independent of metals and glucocorticoid hormones. Proc Natl Acad Sci USA 81: 1053–1056

Ezekowitz RAB, Orkin SH, Newburger PE (1987) Recombinant interferon gamma augments phagocyte superoxide production and X-chronic granulomatous disease gene expression in X-linked variant granulomatous disease. J Clin Invest 80: 1009–1016

Ferris DK, Harel-Bellan A, Morimoto RI, Welch WJ, Farrar WL (1988) Mitogen and lymphokine stimulation of heat shock proteins in T lymphocytes. Proc Natl Acad Sci USA 85: 3850–3854

Granelli-Piperno A, Andrus L, Steinman RM (1986) Lymphokine and nonlymphokine mRNA levels in stimulated human T cells. Kinetics, mitogen requirements and effects of cyclosporin A. J Exp Med 163: 922–937

Halliwell B (1987) Oxidants and human disease: some new concepts. FASEB J 1: 358–364

Helqvist S, Hansen BS, Johannesen J, Andersen HU, Nielsen JH, Nerup J (1989) Interleukin-1 induces new protein formation in isolated rat islets. Acta Endrocrinol (Copenh) 121: 136–140

Hightower LE, White FP (1981) Cellular responses to stress: comparison of a family of 71–73-kilodalton proteins rapidly synthesized in rat tissue slices and canavanine-treated cells in culture. J Cell Physiol 108: 261–275

Jäättelä M, Saksela K, Saksela E (1989) Heat shock protects WEHI-164 target cells from the cytolysis by tumor necrosis factor α and β. Eur J Immunol 19: 1413–1417

Karasawa M, Hosoi J, Hashiba H, Nose K, Tohyama C, Abe E, Suda T, Kurok T (1987) Regulation of metallothionein gene expression by 1α, 25-dihydroxyvitamin D_3 in cultured cells and in mice. Proc Natl Acad Sci USA 84: 8810–8813

Kaur P, Welch Wj, Saklatvala J (1989) Interleukin-1 and tumor necrosis factor increase phosphorylation of the small heat shock protein. Effects in fibroblasts, Hep G2 and U937 cells. FEBS Lett 258: 269–273

Keyse SM, Tyrrell RM (1989) Heme oxygenase is the major 32-kDa stress protein induced in human skin fibroblasts by UVA radiation, hydrogen peroxide, and sodium arserite. Proc Natl Acad Sci USA 86: 99–103

Kingston RE, Baldwin AS, Sharp A (1985) Transcription control by oncogenes. Cell 41: 3–5

Kluger MJ (1986) Is fever beneficial? Yale J Biol Med 59: 89–95

Kluger MJ, Ringler DH, Anver MR (1975) Fever and survival. Science 188: 166–168

Koga T, Wand-Württenberger A, DeBruyn J, M Munk ME, Schoel B, Kaufmann SHE (1989) T cells against a bacterial heat shock protein recognize stressed macrophages. Science 245: 1112–1115

Köller M, Brom C, Brom J, König W (1989) Heat shock induces alterations of lipoxygenase pathway in human polymorphonuclear granulocytes. Prost Leuk Essential Fatty Acids 38: 99–106

Kubo T, Towle CA, Mankin HJ, Treadwell BV (1985) Stress-induced proteins in chondrocytes from patients with osteoarthritis. Arthr Rheumatol 28: 1140–1145

Lederman HM, Brill CR, Murphy PA (1987) Interleukin 1-driven secretion of Interleukin 2 is highly temperature-dependent. J Immunol 138: 3808–3811

Malkinson FD (1980) The heat's on. Arch Dermatol 116: 885–887

Maridonneau-Parini I, Tringale JN, Tauber AI (1986) Identification of distinct activation pathways of the human neutrophil NADPH oxidase. J Immunol 137: 2925–2930

Maridonneau-Parini I, Clerc J, Polla BS (1988) Heat shock inhibits NADPH oxidase in human neutrophils. Biochem Biophys Res Commun 154: 179–186

Morange M, Dubois MF, Bensaude O, Lebon P (1986) Interferon pretreatment lowers threshold for maximal heat shock response in mouse cells. J Cell Physiol 127: 417–422

Müller R (1986) Cellular and viral fos genes: structure, regulation of expression and biological properties of their encoded products. Biochim Biophys Acta 823: 207–225

Nagata K, Saga S, Yamada KM (1986) A major collagen-binding protein a chick embryo fibroblasts is a novel heat shock protein. J Cell Biol 103: 223–229

Polla BS (1988) A role for heat shock proteins in inflammation? Immunol Today 9: 134–137

Polla BS, Healy AM, Amento EP, Krane SM (1986) 1,25-dihydroxyvitamin D3 maintains adherence of human monocytes and protects them from thermal injury. J Clin Invest 77: 1332–1339

Polla BS, Healy AM, Wojno WC, Krane SM (1987a) Hormone 1,25-dihydroxyvitamin D_3 modulates heat shock reponse in monocytes. Am J Physiol 252 (Cell Physiol 21): C640–C649

Polla BS, Healy AM, Byrne M, Krane SM (1987b) 1,25-dihydroxyvitamin D_3 induces collagen binding to the human monocyte line U937. J Clin Invest 80: 962–969

Polla BS, Healy AM, Wojno WC, Krane SM (1987c) Analysis of the heat shock response in human and porcine cells: effects of 1,25-dihydroxyvitamin D_3. In: Calcium regulation and bone metabolism: basic and clinical aspects, vol. 9. Elsevier, Amsterdam, pp 485–490

Polla BS, Kantengwa S, Junod AF (1989) The stress/heat shock response in human alveolar macrophages (abstr). Eur Resp J 2, S5: 350s

Privalle CT, Fridovich I (1987) Induction of superoxide dismutase in Escherichia coli by heat shock. Proc Natl Acad Sci USA 84: 2723–2726

Saga S, Nagata K, Chen W-T, Yamada KM (1987) A pH-dependent function, purification, and intracellular location of a major collagen-binding glycoprotein. J Cell Biol 105: 517–527

St. John T, Gallatin WM, Siegelman M, Smith HT, Fried VA, Weissman IL (1986) Expression cloning of a lymphocyte homing receptor cDNA: ubiquitin is the reactive species. Science 231: 845–849

Taylor MW, Long T, Martinez-Valdez H, Downing J, Zeige G (1984) Induction of gamma-interferon activity by elevated temperatures in human B-lymphoblastoid cell lines. Proc Natl Acad Sci USA 81: 4033–4036

Tumarkin L, Damewood IV GP, Sreevalsan T (1985) Potentiation of thermal njury in mouse cells by interferon. Biochem Biophys Res Commun 128: 179–184

Wong GHW, Elwell JH, Oberley LW, Goeddel DV (1989) Manganous superoxide dismutase is essential for cellular resistance to cytotoxicity of tumor necrosis factor. Cell 58: 923–931.

Heat Shock Proteins as Antigens

Stress-Induced Proteins in Immune Response to Cancer

P. K. Srivastava and R. G. Maki

1 Introduction

Tumor-specific antigens represent molecules which elicit an immune response in a tumor-bearing host. An immune response was classically believed to be elicited only against "nonself" entities; this definition implied that tumor-specific antigens exist on tumors but not on normal tissues. However, a most extensive search has failed to uncover molecules specific to tumors. Instead, comparison of tumors and normal tissues by biochemical methods and by monoclonal antibodies has uncovered (a) antigens which are expressed in tumor cells in larger quantities than in corresponding normal cells, (b) differentiation antigens, which are normally expressed on cells of a developmental lineage distinct from that of the tumor, and (c) oncofetal antigens, which are normally expressed during embryonic development, but not in normal adult tissues (Old 1981). These antigens may serve as useful markers for diagnosis (Sugarbaker 1985; Garrett and Kurtz 1986), prognosis (Kauppila 1989; Benner et al. 1988), or even immunotherapy (Houghton et al. 1985), but none of these antigenic classes are truly tumor specific. Yet the observations which originally necessitated this term,

Department of Pharmacology, Box 1215, Mount Sinai School of Medicine, City University of New York, New York, NY 10029, USA

Current Topics in Microbiology and Immunology, Vol. 167
© Springer-Verlag Berlin · Heidelberg 1991

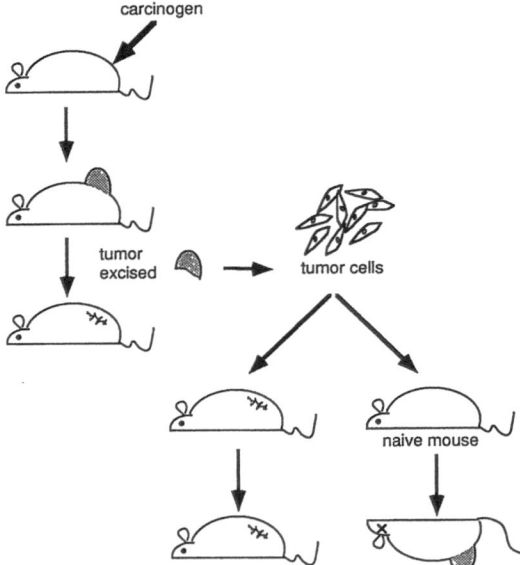

Fig. 1. Tumors elicit immunity in the host of origin. A mouse is injected with a carcinogen, such as MCA, and a tumor develops at the site of injection. The mouse is cured by surgical excision of the tumor. The tumor is passaged by transplantation of trocar fragments to mice of the same genetic pedigree or by in vitro cell culture. The cured mouse is injected with live cells of this tumor and is observed to be resistant to it. In contrast, a naive mouse (of the same pedigree), which has never been exposed to the tumor, succumbs to injection with live tumor cells

i.e., *tumors elicit immune response in the host of origin*, have not only been repeatedly upheld, but substantially expanded.

These observations were first made convincingly in a series of studies with chemically induced tumors of inbred rats and mice (GROSS 1943; FOLEY 1953; BALDWIN 1955; PREHN and MAIN 1957; KLEIN et al. 1960; OLD et al. 1962). These experiments were simple in design (Fig. 1). A tumor is induced in an inbred mouse by injection of a carcinogen such as MCA and is passaged among mice of the same inbred stock. A mouse bearing such a tumor can be cured by surgical excision of the tumor or ligation of tumor blood supply. If the cured mouse is reinjected with the tumor, it is found to be resistant or immune to the tumor; in contrast, naive mice which have never been exposed to the tumor succumb to it. In a surprising development, it was observed that mice are immune only to the tumor to which they had been exposed and not to another tumor, even if the two tumors were induced by the same carcinogen in the same inbred colony of mice and were of the same histological origin (PREHIN and MAIN 1957; BASOMBRIO 1970). This specificity is indeed so exquisite, that even two tumors induced in the same mouse show individually distinct antigenicity (GLOBERSON and FELDMAN 1964). The demonstration of immunity to a tumor by prior exposure to it and the specificity of this immunity have since been extended to a wide variety of sarcomas and carcinomas, induced by a number of carcinogens including ultraviolet irradiation (HOSTETLER and KRIPKE 1987) and in several inbred species including mice, rats, and guinea pigs (for review, see SRIVASTAVA and OLD 1988).

While immunogenic antigens of chemically induced tumors of rodents are generally individually distinct, occasional instances of cross-reactivity have been recorded (COGGIN and ANDERSON 1974; SRIVASTAVA and OLD 1989). Chicken fibrosarcomas usually show cross-reactive immunogenicity (EDELMAN et al. 1983). Also, tumors induced by some DNA viruses show a cross-reactive pattern of immunity: all tumors induced by a given virus possess the same antigenic specificity in tumor rejection (TEVETHIA and BUTEL 1987). Recently, malignant cell lines generated by transfection of oncogenes have been shown to possess mostly cross-reactive and some unique specificities (FREY and LEVINE 1990). The molecular nature of nonviral cross-reactive immunogenic antigens is not known.

The identity of tumor antigens which elicit individually distinct immunity has been the subject of much speculation and little structural analysis (SRIVASTAVA and OLD 1988). A small number of molecules have been biochemically defined (Table 1); of these, some have turned out to be viral antigens expressed on tumor cells (ZBAR et al. 1981; S. LEGRUE and N.R. PELLIS, personal communication), while the relationship of some others to viral antigens is not clear (SATO et al. 1987; RANSOM et al. 1981). An impressive majority of antigens which are clearly unrelated to viral antigens show homology with stress-induced proteins and are discussed here.

2 gp96 (hsp100) Antigens

gp96 antigens were identified by separating fractions of cytosol and plasma membranes of three different MCA-induced BALB/c mouse sarcomas (Meth A, CMS5, and CMS13) by column chromatography and testing individual fractions for their ability to elicit immunity to large tumor challenges of 100 000 or more cells (SRIVASTAVA and OLD 1988; SRIVASTAVA et al. 1986; PALLADINO et al. 1987). The tumor rejection antigen was found to be present in cytosol as well as membranes; interestingly, the immunogenic activity of the cytosol gp96 was masked by a tumor-enhancing activity (TEA) and was not observed until the two activities were separated on a concanavalin A-agarose or DEAE-agarose column (SRIVASTAVA et al. 1986). At least part of the membrane antigen is present on the cell surface, as shown by immunoprecipitation of gp96 from surface radioiodinated cells and by erythrocyte rosetting with anti-gp96 antiserum (see Sect. 6). Immunity elicited by gp96 was found to be specific for the tumor from which gp96 was prepared: mice immunized with Meth A gp96 were resistant to large challenges of Meth A, but not CMS5 sarcoma; conversely, mice immunized with CMS5 gp96 were immune to large challenges of CMS5, but not Meth A. Similar to immunogenicity of the antigens from two UV-induced sarcomas (RANSOM et al. 1981) and the 30-kD antigen from the MCA-induced C3H

Table 1. Tumor antigens identified by their ability to elicit a protective immune response to tumor challenges

Protein	Source	Relationship to:			Reference
		Viral proteins	Hsp's	Other proteins	
p90–100	SV40-transformed cells	SV40 T antigen			ANDERSON et al. 1977
gp70	MCA-induced C3H mouse sarcoma	MuLV gp70 antigen			ZBAR et al. 1981
gp30	MCA-induced C3H mouse sarcoma	MMTV antigen			LEGRUE and PELLIS, unpubl.
gp76	UV-induced mouse sarcoma	?			RANSOM et al. 1981
gp30	Mouse colon carcinoma	?			SATO et al. 1987
gp100	Rat hepatoma		hsp100?		SRIVASTAVA and DAS 1984
gp96	MCA-indued BALB/c mouse sarcoma		hsp100		SRIVASTAVA et al. 1986
p84/86	MCA-induced BALB/c mouse sarcoma		hsp90		ULLRICH et al. 1986
067	Oncogene transfected rat fibroblasts		hsp70		KONNO et al. 1989
B700	B16 mouse melanoma		Albumin		HEARING et al. 1986
p67	Rat histicytoma		Albumin		DESHPANDE et al. unpubl.
p60	Mouse mastocytoma			Unknown	DEPLAEN et al. 1988
p75/82	MCA-induced BALB/c sarcomas			Unknown	DuBois et al. 1982
gp175	MuLV-induced leukemia			Unknown	ROGERS et al. 1984
gp95	A-MuLV-transformed lymphoid cells			Unknown	MACHIDA and KABAT 1983

sarcomas, immunogenicity of Meth A gp96 also appears to be dose restricted; immunization with smaller or larger doses than an optimal dose does not elicit immunity (SRIVASTAVA et al. 1986). Immunity elicited by gp96 is T cell mediated: T lymphocytes from mice immunized with tumor-derived gp96 can adoptively transfer tumor immunity to naive mice and can mediate regression of preexisting tumors in a tumor-specific manner (PALLADINO et al. 1987; SRIVASTAVA et al. 1990).

Rabbit antisera to Meth A derived gp96 were used to determine the distribution of gp96 proteins in normal tissues and tumors; gp96-like proteins were observed to be widely distributed in a range of normal tissues and tumors. In preliminary experiments, gp96 preparations from normal tissues did not elicit immunity to any of the three tumors tested (unpublished results).

The above-mentioned observations suggest that gp96 antigens of normal tissues are different from gp96 derived from tumors and that tumor-derived gp96 molecules also differ from tumor to tumor or are individually distinct. In order to provide a structural definition to this paradigm, gp96 cDNAs and genes have been cloned and characterized from normal tissues and Meth A and CM55 sarcomas (SRIVASTAVA et al. 1987; unpublished results). Three to four genes have been identified in murine as well as human tumors and normal tissues (MAKI et al. 1990). However, no individually distinct tumor-specific genetic alterations have been detected so far. It is conceivable that glycosylation or other post-translational modifications, not evident in the primary sequence, may be responsible for this specificity. However, immunity against these tumors is T cell mediated and there is no precedent for T cell immunity against carbohydrate or lipid determinants.

One possible resolution of this problem may lie in the homology of gp96 with the hsp family. Comparison of gp96 sequences to existing sequences in databases shows extensive homology between gp96, murine endoplasmic reticular protein ERp99/endoplasmin, hamster grp94, and chicken hsp108 cDNAs (KULOMAA et al. 1986; MAZZARELLA and GREEN 1987). Further, a molecule with biochemical properties similar to gp96 has been identified as hsp100 (WELCH et al. 1983). These similarities and our experimental observations that gp96 transcripts are inducible by heat shock (T.-J. CHANG and P.K. SRIVASTAVA, unpublished observations) led us to suggest that gp96 antigens indeed belong to the hsp100 family (MAKI et al. 1990; SRIVASTAVA 1989) (see also last paragraph of the following section for comparison of gp96 and hsp90). The significance of this observation in the elucidation of the structural basis of specific immunity elicited by gp96 is discussed in Sect. 7.

3 p84/86 (hsp90) Antigens

The p84/86 antigens were isolated from Meth A sarcoma by column chromatography of cytosol fractions (ULLRICH et al. 1986), which were assayed by their ability to elicit specific immunity against 20 000 Meth A cells. The antigen was found to consist of two acidic isoforms of 84 and 86 kD present in equimolar amounts. Both isoforms are phosphorylated and do not possess any sugars as determined chemically and by lack of lectin binding. Amino terminal sequences of the p84 and p86 were identical except for short segments which were unique to each isoform. Antigens were purified from the cytosol and immunofluorescence analysis with anti-p84/86 antiserum substantiated a cytosolic localization; a small proportion of p84/86 was seen on the cell surface by immunoelectron microscopy. At the time of identification of p84/86 as a specific tumor immuno-gen, it was difficult to visualize how intracellular antigen could be accessible to the immune system. However, since the demonstration of presentation of internal proteins as peptides by class I MHC antigens (TOWNSEND et al. 1985, 1986), similar to the situation with class II MHC antigens (SHIMONKEVITZ et al. 1983; ALLEN et al. 1984; BABBITT et al. 1985), the distinction between cell surface and internal antigens has become somewhat blurred. Immunity elicited by Meth A p84/86 antigens was tumor specific, even though p84/86-related molecules were found in several antigenically distinct chemically and virally induced tumors.

Recently cDNAs encoding p84/86 antigens have been isolated and sequenced and the corresponding genes have been mapped to mouse chromosomes 2, 12, and 15 (MOORE et al. 1987). Amino acid sequences of mature protein as well as DNA sequence indicate that p84/86 antigens are murine counterparts of the hsp90 family of proteins common to organisms from *E. coli* to man. Similar to the situation with gp96 antigens, no individually distinct or tumor-specific DNA sequence polymorphisms have yet been identified, and the structural basis of specificity of immunogenicity of p84/86 is unclear (S.J. ULLRICH, personal communication).

At this point, it is worthwhile examining the homology between gp96 (hsp100) and p84/86 (hsp90). While overall homology between the two proteins is only 49%, specific regions show as much as 75% homology (Fig. 2). Notable in Fig. 2 are conservation of the relative spacing of highly homologous and

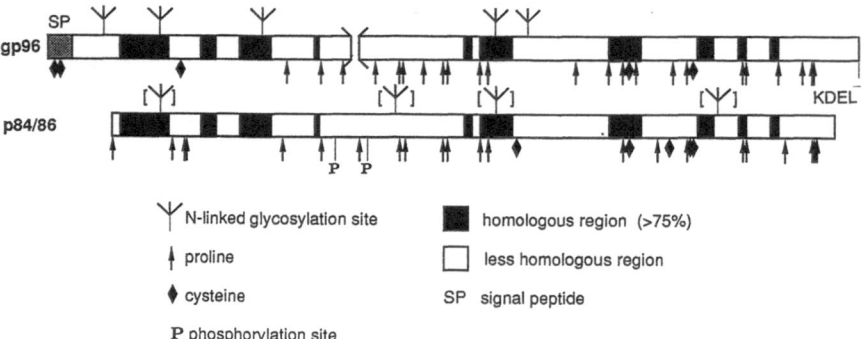

Fig. 2. gp96 (hsp100) and p84/86 (hsp90) molecules share common structural elements

less homologous regions and identical localization of several prolines and two cysteines. The suggestion that the two molecules share a similar three-dimensional conformation has been confirmed by platinum/carbon rotary shadowing and electron microscopy of hsp100 and hsp90 preparations (KOYASU et al. 1986). Prominent differences between the two molecules are the presence of N-linked glycosylation in gp96 and of additional sequences in the 5′ and 3′ regions of gp96 cDNAs. Additional sequences on the 5′ end include the first four exons of gp96, the first of which contains sequerces encoding the hydrophobic signal peptide presumably responsible for cell surface localization of gp96 (SRIVASTAVA 1989). The 3′ end of gp96 contains a lysine–aspartate–glutamate–leucine (KDEL) sequence, which has been associated with proteins retained in the endoplasmic reticulum (PELHAM 1986). This is consistent with a large proportion of a protein closely related or identical to gp96 being present abundantly in the endoplasmic reticulum (MAZZARELLA and GREEN 1987). The question of subcellular localization of gp96 (hsp100), p84/86 (hsp90), and hsp70 molecules is discussed in Sect. 6.

4 p67 (hsp70) Antigen

A protein related to the hsp70 family has been shown to be immunogenic in the case of oncogene-transformed rat fibroblasts (KONNO et al. 1989). Normal rat fetal fibroblasts were transfected with the oncogene H-ras. Two of ten rat transfected clones had detectable levels of a 67-kD protein at the cell surface using a monoclonal antibody (mAb 067) generated from mice immunized with one of these clones. Immunostaining with mAb 067 could be enhanced by heat shock or hydrogen peroxide treatment of these clones, as well as by treatment with reagents that increase cellular cAMP, namely cholera toxin or dibutyryl cAMP. Other (8/10) H-ras-transformed fibroblasts did not express the 67-kD antigen constitutively, but mAb 067 reactivity was induced in these clones by conditions that increase cAMP, though not by heat shock or hydrogen peroxide. Nontransformed fibroblasts and fibroblasts transfected with v-src, polyoma middle T, or c-myc, or transformed with adenovirus type 12 E1A-E1B had no detectable levels of the 67-kD antigen, nor could mAb 067 reactivity be induced by treatment with cholera toxin or dibutyryl cAMP. Monoclonal antibody RPN. 1197 (Amersham), which detects members of the hsp70 family, reacted with the 67-kD antigen, indicating that it is a member of the hsp70 family. No immunological function of this hsp70 homologue has been published. However, preliminary evidence indicates that the two mAb 067-positive clones are rejected in tumor transplantation, even at challenges of more than 10^6 cells, while mAb 067-negative clones grow rapidly, suggesting that this hsp70 homologue may be a tumor rejection antigen in this system. (See Sect. 6 for subcellular localization of hsp70.)

5 Albumin-like Antigens

Albumin is heat shock inducible in the fetal liver (SRINIVAS et al. 1987), although it is constitutively expressed in adult life. This nonclassical stress-inducible protein acts as a tumor antigen in two systems. In mouse melanomas, an albumin-related antigen (B700) elicits cross-reactive'immunity: administration of B700 renders mice resistant to subcutaneous challenges with three melanoma cell lines and decreases the incidence of lung metastases after intravenous challenge with melanoma cells, but does not protect against a sarcoma or a leukemia (HEARING et al. 1986). A corresponding molecule from normal melanocytes (C700) affords no protection against melanomas (KERNEY et al. 1977). However, administration of B700 does not protect mice against all melanomas. S91 mouse melanoma has undetectable levels of B700 at the cell surface, although B700 can be extracted from the cytoplasm. B700 preparations from this melanoma are not protective against challenge with S91 or any other melanoma. However, S91 does contain another antigen distinct from B700 that protects mice specifically against S91 challenge (LAW et al. 1987). Thus there are both more than one type of B700 antigen and more than one type of tumor antigen in mouse melanomas.

N-terminal sequencing of the purified B700 protein provided the first indication of similarity to albumin. Three B700 preparations from different passages of the same melanoma were sequenced and were found to have similar but not identical sequences. Comparison of these sequences to corresponding sequences from albumin showed that between 7 of 15 residues and 11 of 14 residues (depending upon the preparation) of B700 and albumin were identical (MARCHALONIS et al. 1984). Antisera to B700 do not cross-react with albumin from five species, although monoclonal antibodies against B700 recognized mouse serum albumin (TOMITA et al. 1985). Reactivity of antisera to albumin with B700 preparations has not been tested. The sequence variation between different preparations is puzzling and suggests that B700 antigens vary with tumor passage. Comparison of the various B700 preparations with one another and with albumin may be instructive with regard to variability of tumor antigens and their role in escape from immune surveillance.

Immunity to a rat histiocytoma can also be elicited by a 67-kD albumin-like protein (G. DESHPANDE and A. KHAR, personal communication). This protein was identified by an antitumor antiserum elicited by immunization of rats with a spontaneous histiocytoma (KHAR 1986). The serum was observed to have tumor-specific reactivity and did not recognize rat serum albumin. However, antibodies to albumin recognize the p67 tumor antigen and lyse the histiocytoma cells in a complement-dependent cytotoxicity assay. Administration of p67 preparations, but not rat serum albumin, rendered rats resistant to the histiocytoma. It is likely that the syngeneic antiserum is recognizing either a minor alteration of albumin or a low molecular mass moiety which is bound to albumin (see Sect. 7).

6 Cell Surface Localization
of Immunogenic Heat Shock Proteins

Traditionally hsp's have been believed to be located intracellularly—in the cytoplasm, Golgi, ER, nuclei, or nucleoli. However, all immunogenic tumor antigen/hsp's discussed above have been localized, at least partly, to the cell surface. gp96 (hsp100) has been shown to be on cell surface by radio-immunoprecipitation from surface-labeled cells and by erythrocyte rosetting; p84/86 (hsp90) antigens have been demonstrated on the surface by immuno-electron microscopy; the 67-kD hsp70-related antigen appears to be on the surface of transfected cells as tested by positive staining of live cells by the mAb 067 antibody; the albuminoid antigen B700 is also observed on the surface and is secreted by some melanomas. (In each of these antigens, however, tumor antigen/hsp's are also present in the cytosolic fraction.) Also, the recent demonstration that interferon-γ-treated or virally infected macrophages, but not unstressed macrophages, can act as targets of cytotoxic T lymphocytes against hsp65 (KOGA et al. 1989) indicates that hsp65 peptides are present on the cell surface under stressed conditions. Further, antibodies to a peptide binding protein, which cross-reacts with hsp70, have been shown to stain living cells, suggesting a cell surface localization of at least a proportion of hsp70 molecules (LAKEY et al. 1987; VANBUSKIRK et al. 1989). A common denominator of the studies describing a cell surface localization of the various hsp's is the fact that they measure an immunological parameter: eliciting tumor immunity or acting as an immunological target. Therefore, it is tempting to speculate that hsp's mediate immunological functions which necessitate their presence at the cell surface. One such possible function may be their role as accessory molecules in antigen presentation (see Sect. 7). However, presence of hsp's at the cell surface poses the question of a possible mechanism by which they are directed to the cell surface. With the exception of gp96, hsp's do not possess signal peptides. It would appear therefore that they are transported to the cell surface either by some undiscovered signal or by an indirect means, such as those described for yeast mating factor a (KUCHLER et al. 1989), interleukin-1, and fibroblast growth factor.

Another question in this context has to do with the mechanism of anchorage of the hsp's to the cell surface. Low molecular mass hsp's and members of the hsp70 family do not have transmembrane domains; hsp90 and hsp100 possess hydrophobic regions (MAZZARELLA and GREEN 1987), but are not transmem-brane molecules (BOOTH and KOCH 1989). The hsp's also do not possess hydrophobic C-terminal sequences which may anchor them to the surface by a phosphatidylinositol-glycan linkage (LOW and SALTIEL 1988). The questions of sorting of hsp's to the cell surface and their anchoring to the cell surface are difficult to resolve in light of the present concepts of these phenomena.

7 Structural Basis of Specificity of Tumor Immunogenicity: Possible Role of Heat Shock Proteins in Antigen Presentation

The persistent relationship among stress-induced proteins and individual specific antigens of tumors is puzzling. Stress-induced proteins are among the most highly conserved proteins and only a limited number of molecular species has been identified so far: there are as many as nine types of hsp70 (LINDQUIST and CRAIG 1988), three types of hsp90 (MOORE et al. 1987), and probably only one hsp100 (MAKI et al. 1990 and unpublished results) in a given species. In comparison, the transplantation polymorphism of chemically induced tumors is apparently endless (GLOBERSON and FELDMAN 1964). It is possible that stress-induced proteins have hitherto undiscovered mechanisms for generation of diversity.

Another possibility which merits consideration is that gp96 and other stress-induced proteins are not tumor antigens per se, but serve as carriers for other immunogens. Stress-induced proteins have been known to bind to a number of different molecules, such as steroid receptors (KULOMAA et al. 1986), immunoglobulin heavy chain (HAAS and WABL 1983), actin (KOYASU et al. 1986) $pp60^{src}$ (LANKS et al. 1982), p53 (HINDS et al. 1987; PINHASI-KIMHI et al. 1986), and fatty acids (GUIDON and HIGHTOWER 1986). They also help to translocate other molecules (DESHAIES et al. 1988; CHIRICO et al. 1988) and bind to ill-folded or unassembled molecules (KOZUTSUMI et al. 1988). Because of these properties, they have been referred to as chaperonins (ELLIS 1987). In the present context, one can imagine that gp96, p84/86, and hsp70 molecules bind to other molecules, which are immunogenic and copurify with them. However, the gp96 and p84/86 immunogenic antigens identified thus far appear to be homogeneous preparations by several highly sensitive methods of detection, including silver staining and amino terminal sequencing. This would suggest that the copurifying immunogenic entities are low molecular weight substances such as carbohydrates, lipids, or small peptides. The role of carbohydrates and lipids in the immunogenicity of these antigens has been mentioned earlier and seems unlikely in view of the T cell immunity elicited by tumors. A likely candidate for the immunogenic entities are therefore low molecular weight peptides. The recent demonstration that some proteins of the hsp70 family, such as hsc70, BiP/grp78, and hsp70 can bind small peptides with specificity and high affinity (LAKEY et al. 1987; VANBUSKIRK et al. 1989; FLYNN et al. 1989) lends credence to this possibility. In this light, stress-induced proteins appear to be similar to antigen-presenting molecules, such as class I and class II MHC antigens (Fig. 3b). Alternatively, in light of the possible role of hsp's in antigen presentation (PARHAM 1989), stress-induced proteins may facilitate peptide–MHC assembly or reassembly. In this model, the peptides transferred from hsp's to MHC antigens would be the immunogenic entities recognized by T cell receptors (Fig. 3c).

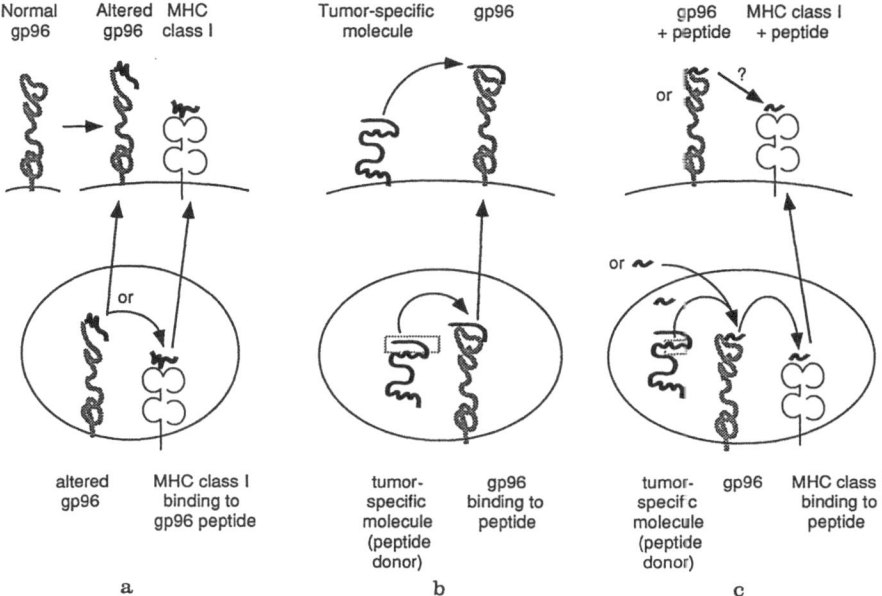

Normal   Altered MHC   Tumor-specific   gp96   gp96   MHC class I
gp96   gp96   class I   molecule   + peptide + peptide

altered   MHC class I   tumor-   gp96   tumor-   gp96   MHC class I
gp96   binding to   specific   binding to   specif c   binding to
            gp96 peptide   molecule   peptide   molecule   peptide
                               (peptide   (peptide
                               donor)   donor)

    a   b   c

Fig. 3. Possible structural bases of specific immunity elicited by gp96 antigens. *Model A:* The gp96 genes show structural variation between normal and tumor cells and among tumors themselves. Immunity is directed against either the intact altered protein or peptides of the processed gp96 antigen. *Model B:* The gp96 antigens are not immunogenic themselves, but become immunogenic as a result of binding of small peptides derived from a distinct antigen. They may be similar to the antigen-presenting molecules of the MHC in this regard. *Model C:* gp96 antigens are accessory to assembly or reassembly of MHC molecules with antigenic peptides derived from a tumor-specific antigen in an intracellular compartment or at the cell surface

These possibilities are consistent with recent developments in the recognition of target antigen by cytotoxic T lymphocytes. It is increasingly clear that not only class II (SHIMONKEVITZ et al. 1983; ALLEN et al. 1984; BABBITT et al. 1985) but also class I molecules (TOWNSEND et al. 1985, 1986) view antigens as peptides and therefore peptides of intracellular antigens can be viewed by the immune system as readily as those of cell surface antigens. Another implication of this new paradigm is that virtually any genetic alteration may lead to the emergence of an immunogenic entity. This has indeed been experimentally demonstrated in the case of tum(−) antigens of P815 mastocytoma (DEPLAEN et al. 1988).

Should the gp96, p84/86, and hsp70 tumor antigens indeed be found to bind peptides, several questions arise. What is the nature of these peptides? Do they derive from a finite set of proteins or are they randomly derived peptides? Should they derive from one or a limited number of proteins, the mechanism of generation of diversity in that family will have to be analyzed. If on the other hand it appears that they are randomly derived, questions about mechanisms of *stability* of this randomness (necessary for specificity) will have to explored. The discovery of a role of stress-induced proteins in tumor immunity has thus uncovered only one layer of this complex phenomenon. The central question in this area—the nature of the antigen—remains, for now, a continuing enigma.

8 Summary

Chemically induced tumors of inbred mice elicit immunity in animals in which the tumors are induced and in other animals of the same inbred stock. The immunity is specific for each tumor: even two tumors induced in one animal with the same carcinogen are not cross-reactive. Immunity to cancer has since been observed in the case of sarcomas and carcinomas induced by a number of chemical and physical carcinogens and in several species, including mice, rats, and guinea pigs. The nature of molecules which mediate immunity to tumors is a central question in cancer immunology. A small number of such molecules have been biochemically defined. Of these, some are viral antigens expressed in tumor cells, while the relationship of some others to viral antigens is unclear. A surprising majority of nonviral tumor antigens have turned out to bear homology with stress-induced proteins. Four families of such molecules are discussed: the gp96 (hsp100) and p84/86 (hsp90) antigens of chemically induced mouse sarcomas, hsp70 antigens of tumors obtained by transfection of normal rat fetal fibroblasts with an H-*ras* oncogene, and the albuminoid antigens of murine melanomas and a rat histiocytoma. (Albumin-like antigens are included among the stress-induced proteins because albumin, though constitutively expressed in adult tissues, is heat shock inducible in fetal liver.) Each of these antigens is a moderately abundant protein, present not only in tumors but also in normal tissues.

Administration of each of these antigen preparations from the tumor, but not from normal tissue, renders the animal immune to challenge with live cells of the tumor from which the antigens are prepared. And yet, no structural differences in the antigens have been observed between normal tissues and tumors. It is suggested that these stress-induced proteins may not be tumor antigens per se, but may be carriers of immunogenic moieties such as short peptides. The stress-induced proteins may therefore serve either as antigen-presenting molecules like the MHC-encoded molecules or as accessory molecules in the presentation of antigens by MHC molecules. The ability of stress-induced proteins to bind to a variety of molecules, including peptides, is consistent with this possibility.

Acknowledgments. The authors thank Drs. Ashok Khar, Stephen LeGrue, Neal Pellis, and Stephen Ullrich for communicating their results prior to publication. The work described here is supported by a National Cancer Institute award (CA44786), a Cancer Research Institute Investigatorship, and an Irma T. Hirschl scholarship. R.G.M. is on leave of absence from Immunology Program, Sloan-Kettering Division, Cornell University Graduate School of Medical Sciences, New York, NY, and receives support from The William Morris Trust.

References

Allen PM, Strydom D, Unanue ER (1984) Processing of lysozyme by macrophages: identification of the determinant recognized by two cell hybridomas. Proc Natl Acad Sci USA 81: 2489–2493

Anderson JL, Martin RG, Chang C, Mora PT, Livingston DM (1977) Nuclear preparations of SV40 transformed cells contain tumor-specific transplantation antigen activity. Virology 76: 420–425

Babbitt B, Allen P, Matsueda G, Haber E, Unanue ER (1985) Binding of immunogenic peptides to Ia histocompatibility molecules. Nature 317: 359–361

Baldwin RW (1955) Immunity to methylcholanthrene-induced tumors in inbred rats following atrophy and regression of the implanter tumours. Br J Cancer 9: 652–657

Basombrío MA (1970) Search for common antigenicities among 25 sarcomas induced by methylcholanthrene. Cancer Res 30: 2458–2462.

Benner SE, Clark GM, McGuire WL (1988) Review: steroid receptors, cellular kinetics, and lymph node status as prognostic factors in breast cancer. Am J Med Sci 296: 59–66

Booth C, Koch GLE (1989) Perturbation of cellular calcium induces secretion of luminal ER proteins. Cell 59: 729–737

Chirico WJ, Waters MG, Blobel G (1988) 70 kD heat shock related proteins stimulate protein translocation into microsomes. Nature 332: 805–810

Coggin JH, Anderson NG (1974) Cancer, differentiation and embryonic antigens: some central problems. Adv Cancer Res 19: 105–164

DePlaen E, Lurquin C, van Pel A, Mariame B, Szikora J-P, Wölfel T, Sibille C, Chomez P, Boon T (1988) Tum⁻ variants of mouse mastocytoma P815. Cloning of the gene of tum⁻ antigen P91A and identification of the tum⁻ mutation. Proc Natl Acad Sci USA 85: 2274–2278

Deshaies RJ, Koch BD, Werner-Washburne M, Craig EA, Schekman R (1988) A sub-family of stress proteins facilitates translocation of secretory and mitochondrial precursor polypeptides. Nature 332: 800–805

DuBois GC, Law LW, Appella E (1983) Purification and biochemical properties of tumor-associated transplantation antigens from methylcholanthrene-induced murine sarcomas. Proc Natl Acad Sci USA 79: 7669–7673

Edelman AS, Xue B, Galton JE, Thorbecke GJ (1983) In vivo and in vitro immune cross-reactivity between dimethylbenzanthracene-induced chicken fibrosarcomas. Fed Proc 42: 1081

Ellis J (1987) Proteins as molecular chaperones. Nature 328: 378–379

Flynn GC, Chappell TG, Rothman JE (1989) Peptide binding and release by proteins implicated as catalysts of protein assembly. Science 245: 385–388

Foley EJ (1953) Antigenic properties of methylcholanthrene-induced tumors in mice of the strain of origin. Cancer Res 13: 835–837

Frey AB, Levine AJ (1990) p53 plus ras transformed rat embryo fibroblasts express tumor-specific transplantation antigen activity which is shared by independently transformed cells. J Virol 63: 5440–5444

Garrett PE, Kurtz SR (1986) Clinical utility of oncofetal proteins and hormones as tumor markers. Med Clin North Am 70: 1295–1306

Globerson A, Feldman M (1964) Antigenic specificity of benzo(a)pyrene induced sarcomas. JNCI 32: 1229–1243

Gross L (1943) Intradermal immunization of C3H mice against a sarcoma that originated in an animal of the same line. Cancer Res 3: 323–326

Guidon PT, Hightower LE (1986) Purification and initial characterization of the 71-kilodalton rat heat-shock protein and its cognate as fatty acid binding proteins. Biochemistry 25: 3231–3239

Haas IG, Wabl M (1983) Immunoglobulin heavy chain binding protein. Nature 306: 387–389

Hearing VJ, Vieira WD, Law LW (1985) Malignant melanoma: cross-reacting (common) tumor rejection antigens. Int J Cancer 35: 403–409

Hearing VJ, Gersten DM, Montague PM, Vieira WS, Galetto G, Law LW (1986) Murine melanoma-specific tumor rejection activity elicited by a purified, melanoma-associated antigen. J Immunol 137: 379–384

Hinds PW, Finlay CA, Frey AB, Levine AJ (1987) Immunological evidence for the association of p53 with a heat shock protein, hsc70, in p53-plus-ras-transformed cell lines. Mol Cell Biol 7: 2863–2869

Hostetler LW, Kripke MI (1987) Origin and significance of transplantation antigens induced on cells transformed by UV irradiation: In: Greene MI, Hamaoka T (eds) Development and recognition of the transformed cell. Plenum, New York, pp 307–330

Houghton AN, Mintzer D, Cordon-Cardo C, Welt S, Fliegel B, Vadhan S, Carswell EA, Melamed MR, Oettgen HF, OLD LJ (1985) Mouse monoclonal IgG3 antibody detecting GD3 ganglioside: a phase I trial in patients with malignant melanoma. Proç Natl Acad Sci USA 82: 1242–1246

Kauppila A (1989) Oestrogen and progestin receptors as prognostic indicators in endometrial cancer: a review of the literature. Acta Oncol 28: 561–566

Kerney SE, Montague PM, Chretien PB, Nicholson JM, Ekel TM, Hearing VJ (1977) Intracellular locatization of tumor-associated antigens in murine and human malignant melanoma. Cancer Res 37: 1519–1524

Khar A (1986) Development and characterization of a rat histiocyte-macrophage tumor line. JNCI 76: 871–879

Klein G, Sjogren HO, Klein E, Hellstrom KE (1960) Demonstration of resistance against methycholanthrene-induced sarcomas in the primary autochthonous host. Cancer Res 20: 1561–1572

Koga T, Wand-Württenburger A, DeBruyn J, Munk ME, Schoel B, Kaufmann SHE (1989) T cells against a bacterial heat shock protein recognize stressed macrophages. Science 245: 1112–1115

Kanno A, Sato N, Yagihashi A, Torigoe T, Cho J, Torimoto K, Hara I, Wada Y, Okubo M, Takahashi N, Kikuchl K (1989) Heat- or stress-inducible transformation-associated cell surface antigen on the activated H-ras oncogene-transfected rat fibroblast. Cancer Res 49: 6578–6582

Koyasu S, Nishida E, Kadowaki T, Matsuzaki F, Iida K, Harada F, Kasuga M, Sakai H, Yahara I (1986) Two mammalian heat shock proteins, HSP90 and HSP100, are actin-binding proteins. Proc Natl Acad Sci USA 83: 8054–8058

Kozutsumi Y, Segal M, Normington K, Gething M-J, Sambrook J (1988) The presence of malfolded proteins in the endoplasmic reticulum signals the induction of glucose-regulated proteins. Nature 332: 462–464

Kuchler K, Sterne RE, Thorner J (1989) Saccharomyces cerevisiae STE6 gene product: a novel pathway for protein export in eukaryotic cells. EMBO J 8: 3973–3984

Kulomaa MS, Weigel NL, Kleinsek DA, Beattie WG, Conneely OM, Marsh C, Zarucki-Schulz T, Schrader WT, O'Malley BW (1986) Amino acid sequence of a chicken heat shock protein derived from the complementary DNA nucleotide sequence. Biochemistry 25: 6244–6251

Lakey EK, Margoliash E, Pierce S (1987) Identification of a peptide binding protein that plays a role in antigen presentation. Proc Natl Acad Sci USA 84: 1659–1663

Lanks KW, Kasambalides EJ, Chinkers M, Brugge JS (1982) A major cytoplasmic glucose regulated protein is associated with the rous sarcoma virus pp60src protein. J Biol Chem 257: 8604–8607

Law LW, Vieira WD, Kameyama K, Hearing VJ (1987) A unique tumor rejection antigen from the S91 murine malignant melanoma. Cancer Res 47: 5841–5845

Lindquist S, Craig EA (1988) The heat shock proteins. Ann Rev Genet 22: 631–677

Low MG, Saltiel AR (1988) Structural and functional roles of glycosylphosphatidylinositol in membranes. Science 239: 268–275

Machida CA, Kabat D (1983) Relationship of surface membranes of lymphoid cells transformed by Abelson murine leukemia virus to tumor rejection. Cancer Res 43: 1275–1281

Maki RG, Old LJ, Srivastava PK (1990) A human homologue of murine tumor rejection antigen gp96: coding region and relationship to stress-induced proteins. Proc Natl Acad Sci USA 87: 5658–5662

Marchalonis JJ, Schwabe C, Gersten DM, Hearing VJ (1984) Amino-terminal variation in melanoma antigens. Biochem Biophys Res Commun 121: 196–202

Mazzarella RA, Green M (1987) ERp99: an abundant conserved glycoprotein of the endoplasmic reticulum is homologous to the 90 kDa heat shock protein (hsp90) and the 94 kDa glucose regulated protein (GRP94). J Biol Chem 262: 8875–8883

Moore SK, Kozak CA, Robinson EA, Ullrich SJ, Appella E (1987) Cloning and nucleotide sequence of the murine hsp84 cDNA and chromosomal assignment of related species. Gene 56: 29–40

Old LJ (1981) Cancer immunology—the search for specificity. Cancer Res 41: 361–375

Old LJ, Boyse EA, Clarke DA, Carswell EA (1962) Antigenic properties of chemically induced tumors. Ann NY Acad Sci 101: 80–106

Palladino MA, Srivastava PK, Oettgen HF, DeLeo AB (1987) Expression of a shared tumor-specific antigen by two chemically induced BALB/c sarcomas. Cancer Res 47: 5074–5079

Parham P (1989) A profitable lesson in heresy. Nature 340: 426–428

Pelham HRB (1986) Speculations on the functions of the major heat shock and glucose-regulated proteins. Cell 46: 959–961

Pinhasi-Kimhi O, Michalovitz D, Ben-Zeev A, Oren M (1986) Specific interaction between the p53 cellular tumor antigen and major heat shock proteins. Nature 320: 182–185

Prehn RT, Main JM (1957) Immunity to methylcholanthrene-induced sarcomas. JNCI 18: 769–778

Ransom J, Schengrund C-L, Bartlett GL (1981) Solubilization and partial characterization of a tumor rejection antigen from an ultraviolet light-induced murine tumor. Int J Cancer 27: 545–554

Rogers MJ, Galetto G, Hearing VJ, Siwarski DF, Law LW (1984) Purification of a glycoprotein bearing a tumor transplantation antigen specific for Friend, Moloney, and Rauscher MuLV-induced tumors. J Immunol 132: 3211–3217

Sato N, Yagihashi A, Okubo M, Torigoe T, Takahashi S, Sato T, Kikuchi K (1987) Characterization of tumor rejection antigen molecules of chemically induced murine colon tumor C-C26. Cancer Res 47: 3147–3151

Shimonkevitz RS, Colon J, Kappler P, Marrack P, Grey HM (1983) Antigen recognition by H-2 restricted T cells. A tryptic ovalbumin peptide that substitudes for processed antigen. J Immunol 133: 2067–2074

Srinivas UK, Revathi CJ, Das MR (1987) Heat-induced expression of albumin during early stages of rat embryo development. Mol Cell Biol 7: 4599–4602

Srivastava PK (1989) Gp96 tumor rejection antigens are cell surface displayed stress proteins. Abstr VII International Congress of Immunology. Gustav Fischer, Stuttgart, p 703

Srivastava PK, Das MR (1984) Serologically unique cell surface antigen of Zajdela ascitic hepatoma is also its tumor associated transplantation antigen. Int J Cancer 33: 417–422

Srivastava PK, Old LJ (1988) Individually distinct transplantation antigens of chemically induced mouse tumors. Immunol Today 9: 78–83

Srivastava PK, Old LJ·(1989) Reply to Coggin's letter. Immunol Today 10: 78

Srivastava PK, DeLeo AB, Old LJ (1986) Tumor rejection antigens of chemically induced tumors of inbred mice. Proc Natl Acad Sci USA 83: 3407–3411

Srivastava PK, Chen YT, Old LJ (19–7) 5' structural analysis of genes encoding polymorphic antigens of chemically induced tumors. Proc Natl Acad SCi USA 84: 3807–3811

Srivastava PK, Maki RG, Zhao JQ, Chang TJ, Heike M (1990) Immunization with soluble gp96 antigens elicits tumor-specific T cell immunity. UCLA Symposia (in press)

Sugarbaker PH (1985) Role of carcinoembryonic antigen assay in the management of cancer. Adv Immun Cancer Ther 1: 167–193

Tevethia SS, Butel JS (1987) SV40 tumor antigen: importance of cell surface localization in transformation and immunological control of neoplasia. In: Greene MI, Hamaoka T (eds) Development and recognition of the transformed cell. Plenum, New York, pp 231–242

Tomita Y, Montague PM, Hearing VJ (1985) Monoclonal antibody production to a B16 melanoma-associated antigen. Int J Cancer 35: 543–547

Townsend A, Gotch F, Davey J (1985) Cytotoxic T cells recognize fragments of the influenza nucleoprotein. Cell 42: 457–467

Townsend A, Rothbard J, Gotch F, Bahadur G, Wraith D, McMichael J (1986) The epitopes of influenza nucleoprotein recognized by cytotoxic T lymphocytes can be defined with short synthetic peptides. Cell 44: 959–968

Ullrich SJ, Robinson EA, Law LW, Willingham W, Appella E (1986) A mouse tumor-specific transplantation antigen is a heat shock related protein. Proc Natl Acad Sci USA 83: 3121–3125

vanBuskirk A, Crump BL, Margoliash E, Pierce SK (1989) A peptide binding protein having a role in antigen presentation is a member of the hsp70 heat shock protein family. J Exp Med 170: 1799–1808

Welch WJ, Garrels JI, Thomas GP, Lin JJ-C, Feramisco JR (1983) Biochemical characterization of the mammalian stress proteins and identification of two stress proteins as glucose- and Ca^{2+}-ionophore-regulated proteins. J Biol Chem 258: 7102–7111

Zbar B, Manohar V, Sugimoto T, Ashley MP, Kato Y, Rappaport P (1981) Immunoprophylaxis of transplantable methylcholanthrene-induced murine fibrosarcomas by immunization with embryo cells expressing endogenous murine leukemia virus antigens. Cancer Res 41: 4499–4507

Heat Shock Proteins as Virulence Factors of Pathogens

R. B. LATHIGRA[1], P. D. BUTCHER[2], T. R. GARBE[1], and D. B. YOUNG[1]

1 Introduction

Infection of a mammal by a bacterial or eukaryotic parasite is a complex and dynamic interaction involving a series of recognition events and phenotypic alterations during which cells of both the host and the microbe undergo a process of mutual recognition and adaptation. The immune response of the host and differentiation associated with the life cycle of protozoan parasites provide classic examples of such adaptation, but recent developments in the field of molecular genetics have also served to focus attention on the changes in gene expression associated with bacterial pathogenicity. Heat shock proteins occupy an intriguing position in the complex web of host–parasite interactions. From the perspective of the infected host, members of the highly conserved heat shock

[1] MRC Tuberculosis and Related Infections Unit, Hammersmith Hospital Du Cane Road, London W12 OHS, UK
[2] Department of Medical Microbiology, St. George's Hospital Medical School, Cranmer Terrace, London SW17 ORE, UK

Current Topics in Microbiology and Immunology, Vol. 167
© Springer-Verlag Berlin · Heidelberg 1991

protein families provide a major stimulus for humoral and cell-mediated immune responses (see chapters by SHINNICK and by WINFIELD and JARJOUR in this volume), and heat shock proteins may also have an important "self-defence" role in cells involved in the inflammatory response (see chapter by POLLA and KANTENGWA in this volume). From the perspective of the pathogen, alteration in heat shock gene transcription has been shown to occur during parasite differentiation, and consideration of the environment encountered by bacterial pathogens—particularly those involved in direct interactions with host phagocytes—suggests that heat shock proteins may also play a role in survival within the host. In this review we will consider heat shock proteins with regard to their possible role in protecting pathogens from microbicidal host functions, and will discuss the relevance of heat shock in relation to in vivo forms of gene regulation.

2 Bacterial Infections: Heat Shock Proteins and Virulence Factors

It has long been recognized that bacteria isolated from infected tissues can show striking differences from the same strains grown in laboratory media, but it is only recently that the powerful new tools of molecular genetics have begun to reveal the molecular mechanisms responsible for such phenotypic alterations (for reviews, see DOUGAN 1989; DIRITA and MEKALANOS 1989; FINLAY and FALKOW 1989; SMITH. 1990). For a wide variety of bacterial pathogens it has been shown that proteins which play an essential role in host–parasite interactions—virulence factors, such as toxins and adhesins—are coordinately regulated at the transcriptional level, and that their expression is induced by environmental stimuli likely to be encountered within the infected host (MILLER et al. 1989). In common with the heat shock proteins, a change in temperature is often employed as a signal for alteration in expression of virulence determinants (MAURELLI 1989) and it is intriguing to draw comparisons, and identify contrasts, between the function and regulation of heat shock proteins and that of well-established virulence factors.

2.1 Function

The multiple gene products which can regulate the complex interaction between a bacterial pathogen and its mammalian host are generally described as virulence factors. This definition encompasses a wide range of molecules, from those which have a direct role in manifestation of overt disease, such as toxins, to those which function by mediating entry or adhesion to host cells, and those which protect the pathogen from attack by the immune system. The term is usually restricted to factors which are specifically involved in the host–parasite interaction, rather than encompassing those which have a general essential function in bacterial survival. Thus the enzymes involved n biosynthesis of aromatic compounds are not described as virulence factors, even though mutation of these genes generally results in attenuation of virulence (HOISETH and STOCKER 1981; ROBERTS et al. 1990).

2.1.1 Are Heat Shock Proteins Virulence Factors?

Two groups of heat shock proteins which have been studied extensively in *E. coli* comprise (a) DnaK [the *E. coli* member of the 70-kilodalton (kD), hsp70, heat shock family] and its functionally associated partners DnaJ and GrpE (GEORGOPOULOS et al. 1989), and (b) GroEL (belonging to the hsp60 family) and its partner, GroES (HEMMINGSEN et al. 1988). Failure to isolate null mutants in the *groE* genes suggest that these proteins are essential for survival (FAYET et al. 1989) and in the case of the DnaK group, although null mutants can be isolated at 30 °C, they have multiple severe growth defects (PAEK and WALKER 1987). It is clear therefore that both of these protein groups perform important functions during normal cell growth of *E. coli* at all temperatures and it can be argued that, as for the enzymes involved in aromatic biosynthesis pathways, heat shock proteins perform an essential "housekeeping" function and are not to be considered as virulence factors. However, it can first be asked whether they have any additional role during infection over and above their normal housekeeping function.

2.1.2 Heat Shock Proteins as Chaperones

Pioneering studies on the role of heat shock proteins during phage infection demonstrated that the DnaK–DnaJ–GrpE group mediates dissociation of a protein complex formed at the origin of replication of phage λ (GEORGOPOULOS et al. 1989), and that GroEL and GroES are required to promote assembly of the λB protein into the head–tail connector of the mature phage (GEORGOPOULOS et al. 1989). Further genetic studies indicate that the GroE proteins interact with DnaA (JENKINS et al. 1986; FAYET et al. 1986) and Ssb-1 (RUBEN et al. 1988) during normal cell growth in *E. coli*, while biochemical evidence suggests a role in translocation of some secreted proteins (BOCHKAREVA et al. 1988). It seems probable that each of these observations reflects a functional role for DnaK and GroEL as "molecular chaperones" or "polypeptide chain binding proteins" in the model proposed by ELLIS (1990) and by ROTHMAN (1989). It is envisaged that, in both eukaryotic and prokaryotic cells, any unfolded regions of polypeptide (newly synthesised chains emerging from the ribosome, for example) are "protected" by reversible association with proteins such as DnaK and GroEL. At an appropriate time, the polypeptide is released and allowed to fold, or assemble with other polypeptide subunits. The polypeptide release signal for the hsp70 class of proteins (including DnaK) seems to be ATP hydrolysis (FLYNN et al.

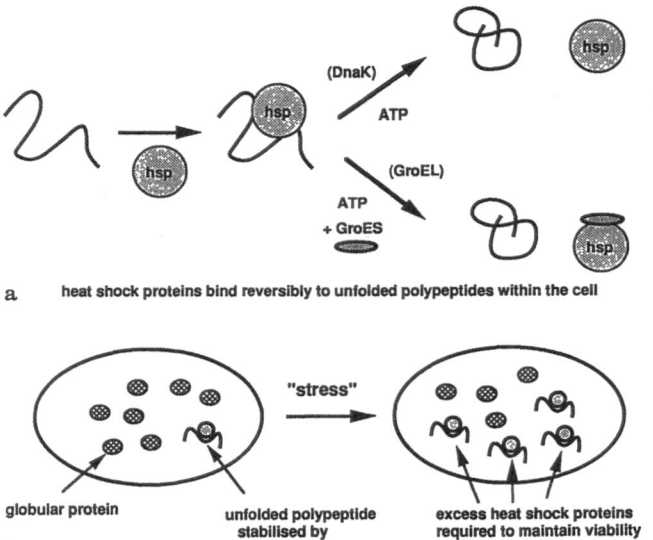

a   heat shock proteins bind reversibly to unfolded polypeptides within the cell

Fig. 1 a, b. Heat shock proteins as chaperones. **a** A model is proposed in which the DnaK and GroEL heat shock proteins bind to unfolded polypeptides within the cell (ROTHMAN 1989; ELLIS 1990). Substrates for this activity would include newly synthesised polypeptides emerging from the ribosome, polypeptides destined for translocation, and globular proteins which have become unfolded following heat shock or other stresses. The heat shock proteins would stabilise the unfolded structure, protecting it from aberrant folding and from non-specific interactions with other intracellular proteins. In response to an appropriate signal—perhaps ATP alone for DnaK, and ATP plus GroES for GroEL—the polypeptide would be released and allowed to fold or assemble into a functional form. **b** In response to heat shock, or other forms of environmental stress, increased levels of heat shock proteins are required in order to cope with the accumulation of partially unfolded or denatured proteins in the cell

1989), while release from GroEL requires both ATP and GroES (GOLOUBINOFF et al. 1989; GEORGOPOULOS and ANG 1990) (Fig. 1). In the light of this model, and bearing in mind the apparent evolutionary pressure for conservation of heat shock proteins, it can be proposed that exposed regions of unfolded polypeptide represent an important threat to cell function and that the ability to increase heat shock protein synthesis in order to neutralise any increase in concentration of unfolded polypeptide is correspondingly essential for cell survival. It can be rationalised therefore that any "stress" which results in increased levels of unfolded polypeptides—by altering patterns of protein synthesis or translocation, or by causing denaturation of proteins within the cell—will be countered by induction of heat shock proteins. If the level of unfolded polypeptide experienced by a pathogen during infection is greater than that with which a free living bacterium has to cope, then it would be possible to consider the chaperones as virulence factors.

2.1.3 Other Heat Shock Proteins

The above discussion refers only to the two groups of heat shock proteins—those linked to DnaK and GroE. Up to 20 heat shock proteins can be identified in *E. coli* by two-dimensional gel electrophoresis (NEIDHARDT and VANBOGELEN 1981) and it is possible that some of the other proteins may play a role in microbial pathogenesis. Two interesting examples, which again focus on the question of denatured polypeptides, are protease La, coded by the *lon* gene (GOFF and GOLDBERG 1985) and the product of the *htrA*, or *degP*, gene (LIPINSKA et al. 1989a). Both of these proteins function as proteases (in the cytoplasm and periplasmic space respectively) and their induction in response to temperature may reflect a role in degradation of polypeptides unfolded as a result of stress. In contrast to the chaperones discussed above, the proteolytic heat shock proteins are required only for survival at high temperature in laboratory culture, with mutant strains being fully viable at 37 °C (LIPINSKA et al. 1989b). Again, it can be suggested that in vivo stress during infection may be countered by induction of such proteins. By using the term "heat shock proteins" to include any protein which is expressed at a higher level following temperature elevation, it is clear that we could extend this category to include several other proteins, such as superoxide dismutase (PRIVALLE and FRIDOVICH 1987) and the temperature-regulated virulence factors (MAURELLI 1989), which have well-established roles in bacterial infection.

2.1.4 Mutations in Heat Shock Genes

On the basis of our current understanding of the function of heat shock proteins we can anticipate that they will play a role in bacterial adaptation during infection, but any attempt to assess the importance of this role reveals a clear need for further experimental evidence. Construction of strains carrying mutations in defined genes provides an important experimental approach to analysing the role of individual proteins in bacterial virulence, and this represents an attractive approach to assessing the significance of heat shock proteins during infection. While it may not be possible to work with null mutants of the DnaK and GroE groups, it is feasible to construct pathogens with temperature sensitive mutations in these genes and it will be of interest to test their ability to survive in vivo. With regard to other heat shock proteins, it has recently been shown that mutations in the *htrA* gene result in attenutation of *Salmonella typhimurium* (G. DOUGAN, personal communication). This exciting finding represents the first experimental evidence of a heat shock protein with a demonstrable role in bacterial virulence.

2.1.5 "Immunovirulence"

Microbial heat shock proteins can stimulate immune responses to both pathogen-specific and autoreactive epitopes and it can be suggested that, by stimulating pathological forms of immune activation, such antigens have a "toxic" effect on the immune system. In the case of chlamydial infection, for example, the GroEL homologue appears to trigger an immune response which is responsible for the tissue damage associated with the disease (MORRISON et al. 1989). Thus, from an immunological rather than a strictly functional standpoint, it may be of interest to consider heat shock proteins as virulence factors (YOUNG et al. 1988b).

2.2 Regulation

The dramatic upregulation of heat shock proteins following temperature elevation has provided an important model for studying gene regulation and has stimulated an extensive body of fundamental research. Many bacterial pathogens experience a rise in temperature on entry into the infected host and, as noted above, temperature often plays a role in regulation of the expression of bacterial virulence. It is of interest to compare the ways in which temperature regulation of heat shock genes and of virulence genes is accomplished.

2.2.1 The Heat Shock Response in *E. coli*

Within 1 min of exposure of *E. coli* to an elevated temperature, synthesis of heat shock proteins increases dramatically and then equilibrates to a new steady state level characteristic of the growth temperature (NEIDHARDT and VANBOGELEN 1987). Heat shock protein synthesis is regulated at the transcriptional level with most of the heat shock genes having a promoter element recognised by *E. coli* RNA polymerase carrying a 32-kD sigma subunit (σ^{32}) (GROSSMAN et al. 1984; SKELLY et al. 1987). The predominant form of RNA polymerase during normal cell growth has a 70-kD sigma subunit but, during heat shock, the level of σ^{32} is elevated by a combination of decreased proteolytic breakdown and increased transcription (STRAUS et al. 1987). Transcription of the *rpoH* gene (which encodes σ^{32}) originates from multiple promoters. Two of the promoters (P1 and P4) are recognised by RNA polymerase carrying the σ^{70} subunit and transcription from these promoters is shut off gradually after a temperature shift to 50 °C. Transcription from a third promoter P3, persists at the high temperature and is mediated by a new sigma factor termed σ^E or σ^{24} (ERICKSON and GROSS 1989). Thus a "cascade" of sigma factors is implicated in the heat shock response. Some of the heat shock genes—such as *htrA*, encoding the periplasmic protease described above—are directly regulated by σ^{24} and lack a σ^{32} consensus promoter (LIPINSKA et al. 1989b). It remains to be determined whether σ^{24} also regulates transcription of other stress-inducible proteins which are not generally included in the list of heat shock proteins.

The mechanisms by which the temperature change is sensed and by which the signal is translated into an alteration in the relative amounts of the different sigma factors remain to be elaborated. In view of the proposed chaperone function of heat shock proteins, it is attractive to propose that the accumulation of unfolded proteins, as suggested by GOFF and GOLDBERG (1985), may be one important component of the regulatory process. By providing an alternative substrate for proteases which may otherwise degrade sigma factors, heat-denatured proteins could have the effect of increasing the intracellular concentration of σ^{32} or σ^{24}. The CIII protein of λ may play such a role in induction of heat shock proteins during phage infection (BAHL et al. 1987).

2.2.2 The Heat Shock Response in Mycobacteria

Almost all of our knowledge about the regulation of the prokaryotic heat shock response is derived from E. coli. In order to determine whether or not E. coli serves as a useful model for studying the heat shock proteins of bacterial pathogens, we have initiated a comparative study of the heat shock response of the human tubercle bacillus, Mycobacterium tuberculosis. Exposure of M. tuberculosis cultures to a change in temperature results in increased synthesis of several proteins identified by metabolic labelling with [^{35}S]methionine and poly-acrylamide gel electrophoresis in the presence of SDS (Fig. 2). In response to moderate heat shock (37 °–42 °C), the 65-kD protein antigen (the mycobacterial GroEL) is seen as a major heat shock protein accounting for more than 10% of the total protein synthesised during the labelling period. At higher temperatures (37 °–48 °C) synthesis of the 65-kD antigen declines while the 71-kD antigen (the DnaK homologue) increases to almost 20% of the [^{35}S]methionine-labelled

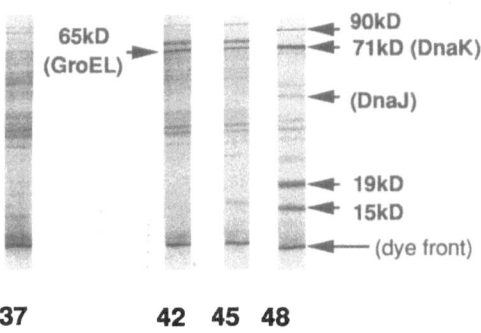

37   42   45   48

incubation temperature °C

Fig. 2. Heat shock response of Mycobacterium tuberculosis. M. tuberculosis H37Rv grown to mid-logarithmic phase at 37 °C was radiolabelled with [^{35}S]methionine during a 1-h incubation at 37, 42, 45 and 48 °C. Bacteria were disrupted by vortexing with glass beads and by boiling in SDS. The pattern of protein synthesis was analysed following separation by SDS-polyacrylamide gel electrophoresis and autoradiography of nitrocellulose blots, which were stained with specific antibodies to localise major mycobacterial antigens. The 65-kD antigen, GroEL, is seen as a major induced protein during mild heat shock, while the 71-kD antigen, DnaK, is prominent also during severe heat shock. The M. tuberculosis GroES homologue, which migrates with an apparent molecular weight of 12 kD, contains no methionine residues and is therefore not seen in this analysis

protein, and unidentified heat shock proteins with molecular weights of 90 kD, 19 kD and 15 kD become prominent (Fig. 2). Several other faint bands are seen, including a 45-kD protein which is recognised by an antiserum raised to the DnaJ homologue of *M. tuberculosis* (LATHIGRA et al. 1988). It is clear that the *M. tuberculosis* GroEL and DnaK proteins are regulated independently at high temperature, suggesting that at least two forms of regulatory mechanism contribute to the heat shock response. Northern blot analysis and prepulsing with rifampicin shows that, as in *E. coli*, the mycobacterial heat shock response is transcriptionally regulated (B. PATEL and P. D. BUTCHER, unpublished results).

The genes for several of the *M. tuberculosis* heat shock proteins have been cloned and sequenced (SHINNICK 1987; LATHIGRA et al. 1988; BAIRD et al. 1988; YOUNG et al. 1988a; R.B. LATHIGRA, unpublished results). The genomic organisation of heat shock genes differs from that in *E. coli* in two respects. The mycobacterial *groEL* and *groES* genes do not form an operon as has been

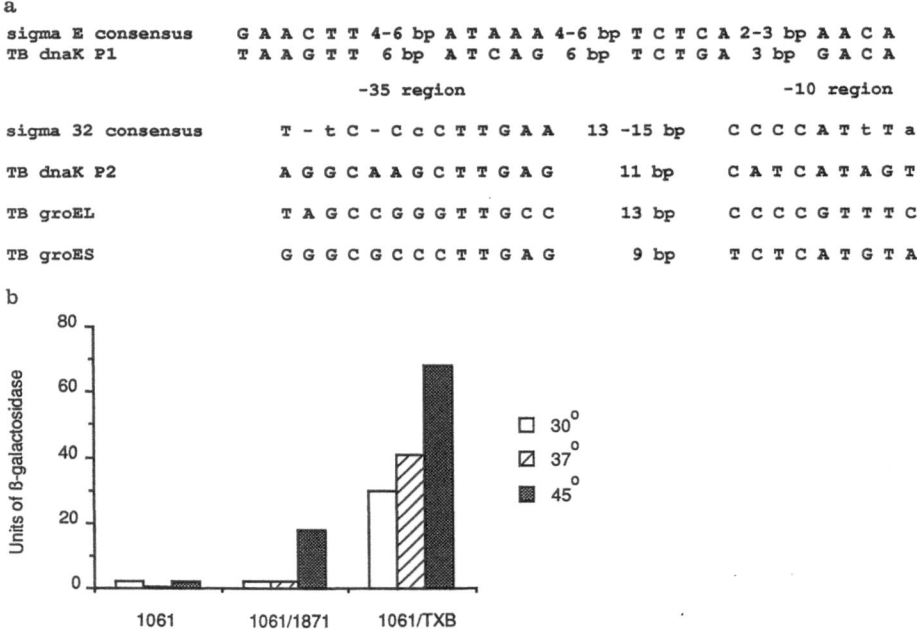

Fig. 3 a, b. Putative promoters of the *Mycobacterium tuberculosis dnaK* gene. **a** Sequence analysis of the promoter region of the *M. tuberculosis dnaK* gene reveals a potential σ^E recognition site (LIPINSKA et al. 1989b)—dnaKP1—and a possible σ^{32} consensus (COWING et al. 1985)—dnaKP2. σ^{32}-like promoters for the *M. tuberculosis groEL* (SHINNICK 1987) and *groES* genes (BAIRD et al. 1988) are also shown. **b** An upstream fragment from the *M. tuberculosis dnaK* gene was fused in frame to the β-galactosidase gene devoid of its promoter and ribosome binding site in the transcription–translation fusion plasmid pMC1871 (CASADABAN et al. 1983). β-Galactosidase activity of the recombinant plasmid, designated 1061/TXB, was measured in *E. coli* strain MC1061 according to the method of MILLER (1972). The mycobacterial promoter(s) stimulated transcription and translation of β-Galactosidase at 30°C, with enhanced activity at higher temperature

described in *E. coli* and other Gram-negative bacteria (HEMMINGSEN et al. 1988; VODKIN and WILLIAMS 1988; MORRISON et al. 1989), and in *M. tuberculosis* the *grpE* gene is closely linked to *dnaK* and *dnaJ* (R. B. LATHIGRA unpublished results). Similarities with *E. coli* are illustrated, however, by the fact that the promoter elements from the mycobacterial heat shock genes are functional in *E. coli* (SHINNICK et al. 1988; R. B. LATHIGRA, unpublished results) and have potential consensus sequences matching those recognised by σ^{32} and σ^{24} (Fig. 3). These results show that there are differences in the arrangement and regulation of heat shock genes between *E. coli* and *M. tuberculosis*, but it has yet to be established whether or not any of these differences relate to the pathogenic lifestyle of the tubercle bacillus.

2.2.3 Temperature and Virulence: Multiple Regulatory Mechanisms

Returning to the topic of virulence factors and their regulation by temperature, is this related to the heat shock response? The mechanisms involved in coordinate regulation of virulence factors have been the subject of several recent incisive reviews (DIRITA and MEKALANOS 1989; FINLAY and FALKOW 1989; GROSS et al. 1989), and we will restrict ourselves to a brief summary of a few systems in which temperature has been shown to act as a signal in virulence gene expression (Table 1).

In two cases sequence analysis has shown that loci involved in virulence regulation are related to a conserved family of bacterial proteins which mediate changes in gene expression in response to a wide range of environmental stimuli. Typically, such "two-component" systems consist of a membrane-spanning protein which can sense changes in the external environment, and a cytoplasmic protein which responds to a signal from the sensor component by direct interaction with DNA to regulate transcription of selected genes (GROSS et al. 1989). The ToxR protein of *Vibrio cholerae* appears to have incorporated both sensor and regular functions into a single protein (MILLER et al. 1987), while the *vir* locus of *Bordetella pertussis* encodes a three-component version of the two-component system (ARICO et al. 1989). In *Yersinia enterocolitica*, temperature regulation of virulence involves the plasmid-borne *virF* gene which encodes a protein with sequence homology to AraC—a DNA-binding protein which controls the arabinose operon in *E. coli* (CORNELIS et al. 1989)—and it has been suggested that the effect of temperature is through a direct modification of the DNA-binding activity of the VirF protein (MAURELLI 1989).

Genes for temperature-regulated virulence factors associated with the ability of *Shigella spp* to invade and colonise eukaryotic cells are carried on a plasmid but their expression is regulated by a repressor protein encoded by the chromosomal gene, *virR* (MAURELLI and SANSONETTI 1988; MAURELLI 1989). A gene analogous to *virR* is present in non-pathogenic *E. coli* (HROMCKYJ and MAURELLI 1989), suggesting that a pre-existing regulatory system may have been adapted for the temperature control of *Shigella* virulence, and DORMAN et al. (1990) have demonstrated that the *virR* gene of *Shigella flexneri* is in fact equivalent to

Table 1. Temperature regulation of bacterial virulence

Pathogen	Virulence factors	Regulatory signals	Regulatory locus	Mechanism
Vibrio cholerae	Toxin Toxin coregulated pilus Accessory colonisation factor	Temperature pH Osmolarity	*toxR*	"Two-component" system
Bordetella pertussis	Toxin Filamentous haemagglutinin Haemolysin	Temperature Mg_2SO_4 Nicotinic acid	*vir* (*bvgA, bvgB, bvgC*)	"Two-component" system
Yersinia spp	Virulence-associated outer membrane proteins (*yop* genes)	Temperature Ca^{2+}	*virF*	DNA-binding protein (AraC family)
Shigella spp	Toxin Invasion plasmid antigens (*ipa* genes) Contact haemolysin	Temperature	*VirR*	Histone-like repressor (supercoiling?)
Escherichia coli	Pilus-adhesin (*pap* genes)	Temperature	*drdX*	Histone-like repressor (supercoiling?)
Listeria monocytogenes	Haemolysin	Temperature	?	?

E. coli osmZ—gene which is important in the environmental control of DNA supercoiling (HIGGINS et al. 1988). In a recent study of temperature regulation of pili-adhesins in uropathogenic *E. coli*, GORANSSON et al. (1990) identified a gene which encodes a repressor protein which is active at low temperature. Cloning of the repressor gene showed that it is probably identical to *virR* and *osmZ*, and that the repressor protein is a 16-kD histone-like polypeptide which had previously been identified as a DNA-binding protein in *E. coli*. It is concluded, therefore, that the overall structure of bacterial chromatin—as influenced by supercoiling and by histone-like binding proteins—plays an important role in the temperature-dependent expression of virulence factors.

In each of these examples the temperatures effective in regulation of virulence have been within the physiologically relevant range. In the case of the haemolysin of *Listeria monocytogenes*, however, it has been reported that gene expression is induced when recently isolated pathogenic strains are exposed to a temperature of 48°C (SOKOLOVIC and GOEBEL 1989). The mechanism of this regulation is unknown, but the experimental conditions are clearly similar to those used to induce maximal heat shock protein synthesis.

Thus, in a brief review of a few selected systems, it can be seen that, while temperature regulation of virulence is a common theme, different mechanisms of regulation have been adopted by different pathogens.

2.2.4 Other Stresses: Oxidative Stress in *Salmonella*

In addition to temperature increase on entering a mammalian host, it is anticipated that bacterial pathogens will be exposed to a variety of other "stress" stimuli. Of particular importance is the oxidative stress associated with exposure to phagocytic cells of the host immune system (HASSET and COHEN 1989). Bacteria present in inflammatory foci and those pathogens which enter host phagocytes will be exposed to a variety of reactive oxygen metabolites, and their ability to survive such exposure may play an important role in the infective process. Are heat shock proteins induced under such circumstances and do they have a critical role in bacterial survival?

Oxidative stress has been characterised in detail by studying exposure of *Salmonella typhimurium* to hydrogen peroxide. During adaptation to hydrogen peroxide, 30 proteins are induced, including heat shock proteins such as the DnaK chaperone (CHRISTMAN et al. 1985; MORGAN et al. 1986). The group of proteins which has been most extensively studied are nine proteins (including DnaK) which are coordinately regulated by the product of the *oxyR* gene. Mutant strains of *S. typhimurium* in which the *oxyR*-regulated genes are constitutively expressed are resistant to killing by hydrogen peroxide, while strains unable to induce the *oxyR* regulon have enhanced hydrogen peroxide sensitivity and, in contrast to the parent strain, are unable to survive in macrophage cultures (MORGAN et al. 1986; FIELDS et al. 1986). This observation provides a basis for suggesting that the DnaK heat shock protein can indeed be regarded as a virulence factor for *S. typhimurium*. Sequence analysis of the *oxyR* gene shows that it codes for a 34-kD protein belonging to a family of bacterial activator proteins (including LysR, CysB and IlvY) with characteristic helix-turn-helix DNA-binding motifs (TAO et al. 1989), but the mechanisms by which hydrogen peroxide alters the expression or function of OxyR remain to be established. It seems, therefore, that although some heat shock proteins are induced in response to hydrogen peroxide stress, this occurs by a mechanism which differs from that of the classic heat shock response.

Hydrogen peroxide represents only one of the reactive oxygen species likely to be produced during the phagocytic respiratory burst and it is of interest to consider the role of heat shock proteins in response to other forms of oxidative stress. Exposure of E. coli to superoxide radicals has been investigated using redox cycling reagents and, while the response overlaps to some extent with oxyR, the GroE heat shock proteins have been implicated in this form of oxidative stress. Different groups have reported this effect to be reflected predominantly either by the GroEL (WALKUP and KOGOMA 1989) or by the GroES protein (GREENBERG and DEMPLE 1989), even though the two genes would seem to form a single operon in E. coli. Synthesis of another heat shock protein, HtpG or C62.5, the bacterial equivalent of the eukaryotic 90-kD, hsp90, family (BARDWELL and CRAIG 1987), is also induced by superoxide radicals (GREENBERG and DEMPLE 1989). Regulatory loci and molecular mechanisms involved in the superoxide-induced stress response remain to be identified.

Study of these simple laboratory models suggests that bacterial adaptation to oxidative stress imposed by host phagocytes is likely to be a complex phenomenon representing the superimposition of different regulatory circuits triggered by different oxidative radicals (HASSET and COHEN 1989). In addition to exposure to oxidative radicals, nutrient limitation may be an additional stress which must be countered during infection and again induction of heat shock proteins may be important in this regard (JENKINS et al. 1988). While it seems probable that the major heat shock proteins will play a role in bacterial adaptation to oxidative stress during infection therefore, the mechanisms involved in the control of their expression may well provide additional complexities to those outlined above in terms of temperature regulation.

3 Parasitic Infections:
Heat Shock Proteins and Differentiation

Parasites have evolved a balance of infection with the host to allow continued host survival with maximum opportunity for horizontal infection. It may be argued that the acquisition of virulence determinants during co-evolution of host and parasite led to the obligate expression of such determinants at different stages of the parasite's life cycle. Hence, phenotypic characteristics of life cycle stages that are involved in the interaction with the host may in this context be regarded as virulence determinants. Many dimorphic pathogen parasites encounter major changes in ambient temperature during their complex life cycles, such as when free living larvae or cysts are ingested by warm-blooded hosts, or when parasites are transferred from a poikilothermic insect vector to a homeothermic mammal. Such changes could be expected to induce heat shock protein synthesis and there is indeed evidence that in eukaryotic parasites the expression of heat shock proteins in response to temperature shifts is an early event in a coordinated programme of gene expression leading to differentiation of life cycle stages (VAN DER PLOEG et al. 1985; NEWPORT et al. 1988). Heat shock protein production in response to increases in temperature have been studied in a variety of parasites, including the insect forms (promastigotes) of *Leishmania* and *Trypanosoma*, *Giardia* trophozoites, yeast stages of *Histoplasma capsulatum* (KEATH et al. 1989), *Candida albicans* (WERNER-WASHBURNE et al. 1989), larval stages of *Brugia*, and *Schistosoma* schistosomulae (see NEWPORT et al. 1988 and YOUNG et al. 1990, for reviews).

In *Leishmania braziliensis*, an increase in heat shock protein production has been correlated with an increase in infectivity and associated with transition from promastigote to amastigote forms (SMEJKAL et al. 1988). *Leishmania mexicana* promastigotes in vitro synthesize two sets of proteins with different induction kinetics by either heat transition from 25° to 37 °C, or by infecting mouse peritoneal macrophages at 35 °C followed by temperature shift to 37 °C (ALCINA and FRESNO 1988). Two hours after temperature shift, 69-, 70-, 83- and 90-kD heat shock proteins were expressed and then switched off, followed 24 h later by a new set of ten proteins characteristic of the amastigote stage and concomitant with transformation to amastigotes. Phagocytosis of promastigotes induced a stress response within 1 h, with production of the same set of heat shock proteins as with the temperature shift. After 24 h at 35 °C, the amastigote protein profile was expressed and correlated with morphological transition. These results suggests that heat shock proteins may function at a defined time in the transformation of the parasite. It seems probable that, rather than heat shock providing a trigger for differentiation, production of heat shock proteins is the first programmed phase of a coordinated sequence of gene expression which defines phase transition. Further recent analysis of *L. major* *hsp70* genes has demonstrated three additional *hsp70* genes, two of which are not temperature regulated but are constitutive for the promastigote stage. The third gene shows differential expression between infective metacyclic and noninfective procyclic promastigotes in the absence of temperature change (SEARLE et al. 1989). This mimics the situation in vivo where differentiation into the infective (virulent) form occurs in response to nutritional deprivation in the insect hind gut just prior to the next blood meal. This differentiation is also accompanied by changes in hsp83 levels (SHAPIRA et al. 1988). The presence of stress proteins prior to transmission may pre-adapt the parasite for stress on encountering the mammalian host. Thus heat shock proteins could be considered as general virulence factors in allowing initial survival at elevated temperatures after invasion and prior to differentiation into a stage-adapted phenotype.

A similar pattern of pre-adaptation by expression of heat shock proteins prior to mammalian infection is seen in *Trypanosoma cruzi*. Epimastigotes show strong induction of heat shock protein synthesis with shut off of other cellular synthesis within 30 min of exposure to heat shock. In metacyclic forms, heat shock proteins are already synthesised at 27 °C with no alteration after heat shock, although remaining protein synthesis is shut off (ALCINA et al. 1988). Over 35 proteins induced in response to heat shock were detected by two-dimensional gel electrophoresis—a considerably higher number than previously recorded (LAWRENCE and ROBERT-GERO 1985). The functions of these proteins are not understood, but a possibility is that some may be involved in pre-adaptation of metacyclic forms for intracellular differentiation in the mammalian vector by expression prior to temperature transition (ALCINA et al. 1988). Regulatory signals for this heat shock protein expression are clearly independent of heat and may, as in *Leishmania* promastigotes, be linked to nutrient deprivation. However, this may not be a general parasitic pre-adaptive response since malarial sporozoites do not express hsp70 even though all the blood stages do (BIANCO et al. 1986).

3.1 Temperature Regulation of Parasite Virulence

There is currently little evidence to suggest the existence of temperature-regulated virulence factors in parasites that are expressed independently of life cycle stage transitions. Several studies have demonstrated temperature-dependent induction of certain proteins that may have a role in parasite virulence.

In *Schistosoma* cercariae, temperature shifts from 23 °C to 37 °C or 42 °C primarily induce two proteins of 58 kD and 60 kD unique to this stage, but not hsp70. These proteins could not be induced by heat or other stress in schistosomulae. The cercariae stage-specific heat shock proteins were induced in the 2 to 3-h period during early transformation and such a high degree of regulation suggests the proteins may be involved in transformation (BLANTON et al. 1987). In contrast, the schistosomulae and adults which do not experience abrupt temperature changes in vivo do not express the 58- and 60-kD proteins, but do express an hsp70 cognate both pre- and post-heat shock. Interestingly, one of the serine proteases secreted during cercarial transformation also has a molecular weight of 60 kD and is suggested to have a role in the release of the cercarial glycocalyx associated with decreased susceptibility to immune attack (McKERROW and DOENHOFF 1988). Thus, these transformation-associated proteases of schistosomes may be temperature-inducible virulence factors.

Another example of temperature-induced factors with potential roles in pathogenicity or virulence is seen in the fungal parasite *Histoplasma capsulatum*. This shows a dimorphic life cycle between a mycelial soil stage and an infective yeast stage that survives within the macrophages of mammalian hosts to produce histoplasmosis. This transition can be induced in vitro by temperature shift to 37 °C of the mycelial stage over a period of several days. Factors involved in this temperature-induced morphogenesis may be considered virulence determinants. This was previously thought to include heat shock proteins since thermotolerance appears to be the key determinant of pathogenicity (LAMBOWITZ et al. 1983) and temperature-sensitive mutants with reduced virulence showed reduced levels of expression of hsp70 at lower temperatures than the virulent thermotolerant strain (CARUSO et al. 1987). However, a yeast phase specific (YPS) protein, YPS-3, has recently been detected 24 h after heat shock in virulent strains of *Histoplasma* but was either expressed after a longer period of 3 days in strains of intermediate virulence or not expressed at all in avirulent strains. YPS-3 was not required for transition but was related to virulence (KEATH et al. 1989). During this heat shock induced transition the *hsp70* genes were expressed within 1–3 h in all strains. Regulation of hsp70 and YPS-3 is clearly different and their roles in phase transition remain unknown.

The opportunistic fungal pathogen *Candida albicans* grows as a yeast form on epithelial cell surfaces and tissue invasion may often be initiated by germ tube formation. This transition when induced in vitro by temperature shifts from 23 °C to 37 °C evokes the synthesis of 14 heat shock proteins. The fact that 12 of the 14 heat shock proteins formed by a potentially pathogenic strain were also formed by a germination-defective variant makes it unlikely that they play a role in regulation of germination (ZEUTHEN and HOWARD 1989). This supposition is further supported by the fact that a temperature shift does not take place before *Candida* tissue invasion in vivo as it does in dimorphic pathogens such as *Histoplasma*.

4 Summary and Concluding Remarks

We have discussed four aspects which seem to provide a link between heat shock proteins and microbial virulence:

1. The functions ascribed to heat shock proteins—in protecting the cell from the potentially disruptive effects of unfolded peptides—suggest that they may be important in enhancing survival of pathogens during infection. It remains to be determined, however, whether the different types of "stress" inherent in a pathogenic lifestyle as compared to that of free living bacteria results in any significant qualitative or quantitative differences in the requirement for heat shock proteins.
2. Temperature regulation of gene expression is a common theme shared by heat shock proteins and established virulence factors. Although temperature

is a common signal, a variety of mechanisms are employed for translation of this signal into an alteration in gene expression (including changes in sigma factors, induction of trans-activating DNA-binding proteins, and changes in the physical structure of bacterial chromatin), with the classic heat shock response representing only one of several regulatory circuits. Since the primary sensor of the temperature change has yet to be clearly established, however, it remains possible that there is some form of secondary signal which is shared by the different transduction systems.

3. Induction of heat shock proteins in response to oxidative stress may be important for pathogens which are exposed to host phagocytes. Regulation of heat shock gene expression in response to oxidative stress differs from the heat shock response in that it involves selective induction of different heat shock protein subsets in response to different oxidative radicals. It is probable that the in vivo stress response of a bacterial pathogen engulfed by a host phagocyte will involve simultaneous induction of several overlapping stress regulons and, while it seems unlikely that this could occur without including the major heat shock proteins, the regulatory mechanisms involved remain to be established.

4. Heat shock gene expression is associated with the early stages of parasite differentiation. This induction is probably part of a coordinated programme of differential gene expression, rather than a simple response to temperature change, but it is possible that in some parasites prior induction of heat shock proteins has a protective effect during the initial stages of transfer of a warm-blooded host.

While heat shock proteins may not prove to be the single key to understanding specific aspects of virulence, it is probable that further study of the function and regulation of these ubiquitous molecules will provide a means to tease apart some of the molecular strands involved in the complex interaction between microbial parasites and their mammalian hosts.

BUCHMEIER and HEFFRON (1990) have demonstrated that the major heat shock proteins DnaK and GroEL are indeed included among the proteins induced following phagocytosis of Salmonella (Science 248: 730–732)

References

Alcina A, Fresno M (1988) Early and late heat induced proteins during Leishmania mexicana transformation. Biochem Biophys Res Comm 156: 1360–1367

Alcina A, Urzainqui A, Carrsco L (1988) The heat shock response in Trypanosoma cruzi. Eur J Biochem 172: 121–127

Arico B, Miller JF, Roy C, Stibitz S, Monack D, Falkow S, Gross R, Rappuoli R (1989) Sequence required for expression of Bordetella pertussis virulence factors share homology with prokaryotic signal transduction proteins. Proc Natl Acad Sci USA 86: 6671–6675

Bahl H, Echols H, Straus D, Court D, Crowl R, Georgopoulos CP (1987) Induction of the heat shock response in *Escherichia coli* through stabilization of σ^{32} by the phage λ cIII protein. Genes Dev 1: 57–64

Baird PN, Hall LMC, Coates ARM (1988) A major antigen from *Mycobacterium tuberculosis* is homologous to the heat shock protein GroES from *E. coli* and the *htpA* gene product of *Coxiella burnetti*. Nucleic Acids Res 16: 9047

Bardwell JCA, Craig EA (1987) Eukaryotic Mr 83,000 heat shock protein has a homologue in *Escherichia coli*. Proc Natl Acad Sci USA 84: 5177–5181

Bianco AE, Favaloro JM, Burkof TR, Culvenor JG, Crewther PE, Brown GV, Anders RF, Coppel RL, Kemp DJ (1986) A repetitive antigen of *Plasmodium falciparum* that is homologous to heat shock protein 70 of *Drosophila melanogaster*. Proc Natl Acad Sci USA 84: 8713–8717

Blanton R, Loula EC, Parker J (1987) Two heat shock induced proteins are associated with transformation of *Schistosoma mansoni* cercariae to schistosomula. Proc Nat. Acad Sci USA 84: 9011–9014

Bochkareva ES, Lissin NM, Girshovich AS (1988). Transient association of newly synthesized unfolded proteins with the heat shock GroEL protein. Nature 336: 254–257

Caruso M, Sacco M, Medoff G, Maresca B (1987) Heat shock 70 gene is differentially expressed in *Histoplasma capsulatum* strains with different levels of thermotolerance and pathogenicity. Mol Microbiol 1: 151–158

Casadaban MJ, Martinez-Arias A, Shapira SK, Chou J (1983) β-Galactosidase gene fusions for analyzing gene expression in *Escherichia coli* and yeast. Methods Enzymol 100: 293–308

Christman MF, Morgan RW, Jacobson FS, Ames BN (1985) Positive control of a regulon for defences against oxidative stress and some heat shock proteins in *Salmonella typhimurium*. Cell 41: 753–762

Cornelis G, Sluiters C, Lambert de Rouvroit C, Michiels T (1989) Homology between VirF, the transcriptional activator of the *Yersinia* virulence regulon, and AraC, the *Escherichia coli* arabinose operon regulator. J Bacteriol 171: 254–262

Cowing DW, Bardwell JCA, Craig EA, Woolford C, Hendrix RW, Gross CA (1985) Consensus sequence for *Escherichia coli* heat shock gene promoters. Proc Natl Acad Sci USA 82: 2679–2683

DiRita VJ, Mekalanos JJ (1989) Genetic regulation of bacterial virulence. Annu Rev Genet 23: 455–482

Dorman CJ, Bhrian NN, Higgins CF (1990) DNA supercoiling and environmental regulation of virulence gene expression in *Shigella flexneri*. Nature 344: 789–792

Dougan G (1989) Molecular characterization of bacterial virulence factors and the consequence for vaccine design. J Gen Microbiol 135: 1397–1406

Ellis RJ (1990) The molecular chaperone concept. Semin Cell Biol 1: 1–9

Erickson JW, Gross CA (1989) Identification of the σ^{E} subunit of *Escherichia coli* RNA polymerase: a second alternate sigma factor involved in high-temperature gene expression. Genes Dev 3: 1462–1471

Fayet O, Louarn J-M, Georgopoulos C (1986) Suppression of the *Escherichia coli dnaA46* mutation by amplification of the *groES* and *groEL* genes. Mol Gen Genet 202: 435–445

Fayet O, Ziegelhoffer T, Georgopoulos C (1989) The *groES* and *groEL* heat shock genes of *Escherichia coli* are essential for bacterial growth at all temperatures. J Bacteriol 171: 1379–1385

Fields PI, Swanson RV, Haidaris CG, Heffron F (1986) Mutants of *Salmonella typhimurium* that cannot survive within the macrophage are avirulent. Proc Natl Acad Sci USA 83: 5189–5193

Finlay BB, Falkow S (1989) Common themes in microbial pathogenicity. Microbiol Rev 53: 210–230

Flynn GC, Chappell TG, Rothman JE (1989) Peptide binding and release by proteins implicated as catalysts of protein assembly. Science 245: 385–390

Georgopoulos C, Ang D (1990) The *Escherichia coli groE* chaperonins. Semin Cell Biol 1: 19–25

Georgopoulos C, Tilly K, Ang D, Chandrasekhar GN, Fayet O, Spence J, Ziegelhoffer T, Liberek K, Zylicz M (1989) The role of *Escherichia coli* heat shock proteins in bacteriophage lambda growth. In: Pardue ML, Feramisco JR, Lindquist S (eds) Stress-induced proteins. Alan R Liss, New York.

Goff SA, Goldberg AL (1985) Production of abnormal proteins in *E. coli* stimulates transcription of *lon* and other heat shock genes. Cell 41: 587–595

Goloubinoff P, Christeller JT, Gatenby AA, Lorimer G (1989) Reconstitution of active dimeric ribulose biphosphate carboxylase from an unfolded state depends on two chaperonin proteins and MgATP. Nature 342: 884–889

Goransson M, Sonden B, Nilsson P, Dagberg B, Forsman K, Emanuelsson K, Uhlin BE (1990) Transcriptional silencing and thermoregulation of gene expression in *Escherichia coli*. Nature 344: 682–685

Greenberg JT, Demple B (1989) A global response induced in *Escherichia coli* by redox-cycling agents overlaps with that induced by peroxide stress. J Bacteriol 171: 3933–3939

Gross R, Arico B, Rappuoli R (1989) Families of bacterial signal-transducing proteins. Mol Microbiol 3: 1661–1667

Grossman AD, Erickson JW, Gross CA (1984) The *htpR* gene product of *E. coli* is a sigma factor for heat-shock promoters. Cell 38: 383–390

Hasset DJ, Cohen MS (1989) Bacterial adaptation to oxidative stress: implications for pathogenesis and interaction with phagocytic cells. FASEB J 3: 2574–2582

Hemmingsen SM, Woolford C, van der Vies SM, Tilly K, Dennis DT, Georgopoulos CP, Hendrix RW, Ellis RJ (1988) Homologous plant and bacterial proteins chaperone oligomeric protein assembly. Nature 333: 330–334

Higgins CF, Dorman CJ, Stirling DA, Waddell L, Booth IR, May G, Bretner E (1988) A physiological role for DNA supercoiling in the osmotic regulation of gene expression in *S. typhimurium* and *E. coli*. Cell 52: 569–584

Hoiseth SK, Stocker BAD (1981) Aromatic-dependent *Salmonella typhimurium* are non-virulent and effective live vaccines. Nature 291: 238–239

Hromckyj AE, Maurelli AT (1989) Identification of an *Escherichia coli* gene homologous to *virR*, a regulator of *Shigella* virulence. J Bacteriol 171: 2879–2881

Jenkins AJ, March JB, Oliver IR, Masters M (1986) A DNA fragment containing the *groE* genes can suppress mutations in the *Escherichia coli dnaA* gene. Mol Gen Genet 202: 446–454

Jenkins DE, Schultz JE, Matin A (1988) Starvation-induced cross protection against heat or H_2O_2 challenge in *Escherichia coli*. J Bacteriol 170: 3910–3914

Keath EJ, Painter AA, Kobayashi GS, Medoff G (1989) Variable expression of a yeast phase specific gene in *Histoplasma capsulatum* strains differing in thermotolerance and virulence. Infect Immun 57: 1384–1390

Lambowitz AM, Kobayashi GS, Painter A, Medoff G (1983) Possible relationship of morphogenesis in pathogenic fungus *Histoplasma capsulatum* to heat shock response. Nature 303: 806–808

Lathigra RB, Young DB, Sweetser D, Young RA (1988) A gene from *Mycobacterium tuberculosis* which is homologous to the DnaJ heat shock protein of *Escherichia coli*. Nucleic Acids Res 16: 1636

Lawrence F, Robert-Gero M (1985) Induction of heat shock and stress proteins in promastigotes of three *Leishmania* species. Proc Natl Acad Sci USA 82: 4414–4417

Lipinska B, Fayet O, Baird L, Georgopoulos CP (1989a) Identification, characterization and mapping of the *Escherichia coli htrA* gene, whose product is essential for bacterial growth at elevated temperatures. J Bacteriol 171: 1574–1584

Lipinska B, Sharma S, Georgopoulos CP (1989b) Sequence analysis of the *htrA* gene in *Escherichia coli*: a σ^{32} independent mechanism of heat-inducible transcription. Nucleic Acids Res 16: 10053–10057

Maurelli AT (1989) Temperature regulation of virulence genes in pathogenic bacteria: a general strategy for human pathogens? Microbial Pathogenesis 7: 1–10

Maurelli AT, Sansonetti PJ (1988) Identification of a chromosomal gene controlling temperature regulated expression of *Shigella* virulence. Proc Natl Acad Sci USA 85: 2820–2824

McKerrow JH, Doenhoff MJ (1988) Schistosome proteases. Parasitol Today 4: 334–340

Miller JF, Mekalanos JJ, Falkow S (1989) Coordinate regulation and sensory transduction in the control of bacterial virulence. Science 243: 916–922

Miller JH (1972) Experiments in molecular genetics. Cold Spring Harbor Laboratory, Cold Spring Harbor, New York

Miller VL, Taylor RK, Mekalanos JJ (1987) Cholera toxin transcriptional activator ToxR is a transmembrane DNA binding protein. Cell 48: 271–279

Morgan RW, Christman MF, Jacobson FS, Storz G, Ames BN (1986) Hydrogen peroxide-inducible proteins in *Salmonella typhimurium* overlap with heat shock and other stress proteins. Proc Natl Acad Sci USA 83: 8059–8063

Morrison RP, Belland RJ, Lyng K, Caldwell HD (1989) Chlamydial disease pathogenesis. The 57-kD chlamydial hypersensitivity antigen is a stress response protein. J Exp Med 170: 1271–1283

Neidhardt FC, VanBogelen RA (1981) Positive regulatory gene for temperature-controlled proteins in *Escherichia coli*. Biochem Biophys Res Commun 100: 894–900

Neidhardt FC, VanBogelen RA (1987) The heat shock response. In Neidhardt FC (ed) *Escherichia coli* and *Salmonella typhimurium*. American Society for Microbiol, Washington D.C.

Newport G, Culpepper J, Agabian N (1988) Parasite heat shock proteins. Parasitol Today 4: 306–312

Paek K-H, Walker GC (1987) *Escherichia coli dnaK* null mutants are inviable at high temperature. J Bacteriol 169: 283–290

Privalle CT, Fridovich I (1987) Induction of superoxide dismutase in *Escherichia coli* by heat shock. Proc Natl Acad Sci USA 84: 2723–2726

Roberts M, Maskell D, Novotny P, Dougan G (1990) Construction and characterization in vivo of *Bordetella pertussis aroA* mutants. Infect Immun 58: 732–739

Rothman JE (1989) Polypeptide chain binding proteins: catalysts of protein folding and related processes in cells. Cell 59: 591–601

Ruben SM, van Den Brink-Webb SE, Rein DE, Meyer RR (1988) Suppression of the *Escherichia coli ssb-1* mutation by an allele of *groEL*. Proc Natl Acad Sci USA 85: 3767–3771

Searle S, Campos AJ, Coulson RM, Spithill TW, Smith DF (1989) A family of heat shock protein 70 related genes are expressed in the promastigote of *Leishmania major*. Nucleic Acids Res 17: 5081–5095

Shapira M, McEwen JG, Jaffe CL (1988) Temperature effects on molecular processes which lead to stage differentiation in *Leishmania*. EMBO J 7: 2895–2901

Shinnick TM (1987) The 65 kDa antigen of *Mycobacterium tuberculosis*. J Bacteriol 169: 1080–1088

Shinnick TM, Vodkin MH, Williams JL (1988) The *Mycobacterium tuberculosis* 65 kDa antigen is a heat shock protein which corresponds to common antigen and to the *E. coli* GroEL protein. Infect Immun 56: 446–451

Skelly S, Coleman T, Wu C-F, Brot N, Weissbach H (1987) Correlation between 32-kDa σ factor levels and in vitro expression of *Escherichia coli* heat shock genes. Proc Natl Acad Sci USA 84: 8365–8369

Smejkal RM, Wiff R, Olenick JG (1988) *Leishmaniasis braziliensis panamensis*: increased infectivity resulting from heat shock. Exp Parasitol 65: 1–9

Smith H (1990) Pathogenicity and the microbe in vivo. J Gen Microbiol 136: 377–383

Sokolovic Z, Goebel W (1989) Synthesis of listeriolysin in *Listeria monocytogenes* under heat shock conditions. Infect Immun 57: 295–298

Straus DB, Walter WA, Gross CA (1987) The heat shock response of *Escherichia coli* is regulated by changes in the concentration of σ^{32}. Nature 329: 348–351

Tao K, Makino K, Yonei S, Nakata A, Shinagawa H (1989) Molecular cloning and nucleotide sequencing of *oxyR*, the positive regulatory gene of a regulon for adaptive response to oxidative stress in *Escherichia coli*: homologies between OxyR protein and a family of bacterial activator proteins. Mol Gen Genet 218: 371–376

van der Ploeg LK, Giannini SH, Cantor CR (1985) Heat shock genes: regulatory role for differentiation in parasitic protozoa. Science 228: 1443–1445

Vodkin MH, Williams JC (1988) A heat shock operon in *Coxiella burnetti* produces a major antigen homologous to a protein in both mycobacteria and *Escherichia coli*. J Bacteriol 170: 1227–1234

Walkup LKB, Kogoma T (1989) *Escherichia coli* proteins inducible by oxidative stress mediated by the superoxide radical. J Bacteriol 171: 1476–1484

Werner-Washburne M, Becker J, Kosic-Smithers J, Craig EA (1989) Yeast Hsp70 RNA levels vary in response to the physiological status of the cell. J Bacteriol 171: 2680–2688

Young D, Lathigra R, Hendrix R, Sweetser D, Young RA (1988a) Stress proteins are major immune targets in leprosy and tuberculosis. Proc Natl Acad Sci USA 85: 4267–4270

Young DB, Mehlert A, Bal V, Mendez-Samperio P, Ivanyi J, Lamb JR (1988b) Stress proteins and the immune response to mycobacteria—antigens as virulence factors? Antonie van Leeuwenhoek J Microbiol 54: 431–439

Young DB, Mehlert A, Smith DF (1990) Stress proteins and infectious diseases. In: Morimoto R, Tissieres A, Georgopoulos C (eds) Stress proteins in biology and medicine. Cold Spring Harbor, New York (in press)

Zeuthen ML, Howard DH (1989) Thermotolerance and the heat shock response in *Candida albicans*. J Gen Microbiol 135: 2509–2518

Heat Shock Proteins as Antigens of Bacterial and Parasitic Pathogens

T. M. SHINNICK

1 Introduction

The first clues as to the immunoreactivity of heat shock proteins came from the work of BIANCO et al. (1986). They showed that the amino acid sequence of a major antigen of the malaria parasite had 70% identity to the sequence of a 70-kD heat shock protein of *Drosophila*. Interest in the antigenicity of heat shock proteins was further stimulated by reports that a highly immunoreactive 65-kD protein of mycobacteria had homology with a 60-kD heat shock protein of *Escherichia coli* (SHINNICK et al. 1988; THOLE et al. 1988; YOUNG et al. 1988a). This protein was also shown to be related to a strongly immunoreactive antigen found in virtually all gram-negative bacteria (SHINNICK et al. 1988; THOLE et al. 1988).

It is now realized that heat shock proteins play a role in the immune response to many bacterial and parasitic pathogens, including the etiological agents of malaria, schistosomiasis, trypanosomiasis, filariasis, syphilis, tuberculosis, leprosy, legionnaires' disease, Lyme disease, Q fever, and a number of other bacterial infections. In this chapter, the studies that have identified heat shock proteins as antigens will be reviewed. The implications of this immunoreactivity for the pathogen and the host are discussed elsewhere in this volume. Additional

Hansen Disease Laboratory, Division of Bacterial Diseases, Centers for Disease Control, Atlanta, Georgia 30333, USA

Current Topics in Microbiology and Immunology, Vol. 167
© Springer-Verlag Berlin · Heidelberg 1991

reviews on the immune response to heat shock proteins may be found in D. YOUNG et al. (1988b, 1989), DUBOIS (1989), PALLEN (1989), WINFIELD (1989), YOUNG and ELLIOT (1989), and R. YOUNG (1990).

Immunoreactive proteins have been identified that show homology with members of five families of heat shock proteins; the hsp90, hsp70, hsp60 or GroEL, low molecular weight, and GroES families. A family is composed of similarly sized proteins that display sequence identity with one another. A family is named according to the size of the proteins in kilodaltons, e.g., the hsp70 family contains *heat* *shock* *proteins* of ~ 70 kD. For discussions of the structure, sequences, expression, homologies, and functions of the heat shock proteins, the reader is referred to several recent reviews (LINDQUIST 1986; LINDQUIST and CRAIG 1988; WELCH et al. 1989) as well as other chapters in this volume.

2 The hsp90 Family

Members of the hsp90 family range in size from 83 to 90 kD and include hsp83 of *Drosophila melanogaster* and hsp90 of *Saccharomyces cerevisiae* (reviewed in LINDQUIST 1986; LINDQUIST and CRAIG 1988). The *E. coli* homologue migrates with an apparent size of 62.5 kD. Within eukaryotes, members of this family display impressive amino acid sequence homology, with an overall identity of 60% and short regions that display > 90% identity. The *E. coli* homologue shows ~ 36% identity with the *Drosophila* hsp83. Another stress-response protein, grp94, is also part of this family. The *hsp90* genes can be found in a single copy or as a multigene family in eukaryotic genomes. The expression of individual *hsp90* genes can be constitutive or can be induced by heat shock, stress, or glucose deprivation. hsp90 proteins are abundant cytoplasmic proteins in most cells grown under normal conditions. No enzymatic function has been described for these proteins, although they have been shown to interact with protein kinases and steroid receptors and may play some role in the transport or regulation of these proteins (reviewed in LANGER and NEUPERT, this volume).

The hsp90 proteins of three parasites, *Trypanosoma cruzi, Schistosoma mansoni*, and *Plasmodium falciparum*, have been reported to be antigenic. No bacterial antigens have been described with homology to members of this family.

DRAGON et al. (1987) isolated clones from a cDNA library of *T. cruzi* mRNA that reacted with sera from persons with Chagas' disease. One clone was shown to react with the patient sera and encode part of an 85-kD protein. Sequence analysis of this cDNA clone and parts of a clone containing the genomic copy of this gene revealed that the amino acid sequence of the *T. cruzi* protein had 60%–65% identity with the corresponding regions of hsp83 of *Drosophila* and hsp90 of yeast. The *T. cruzi* genes are found in a cluster of six to ten repeats and are expressed constitutively in both epimastigotes and trypomastigotes. Incidentally, this report appears to be the first to raise the possibility that the immune response to conserved regions of heat shock proteins may play a role in the autoimmune consequences of an infection such as is occasionally seen in Chagas' disease.

Sera of persons infected with *S. mansoni* react with an 86-kD antigen that is thought to be exposed on the surface of schistosomulae (TAYLOR et al. 1984). A cDNA clone that hybridizes to the mRNA encoding this protein was isolated and shown to contain the carboxy terminal portion of its open reading frame (JOHNSON et al. 1989). The deduced 442 amino acid sequence has 59% identity with yeast hsp90 and 64% identity with *Drosophila* hsp83. The *S. mansoni* gene appears to belong to a multigene family, and its expression is heat inducible (JOHNSON et al. 1989).

In squirrel monkeys, protection from infection with *P. falciparum* correlates with the presence of antibodies to proteins of 72 and 90 kD (DUBOIS et al. 1984; JENDOUBI et al. 1985). To isolate cDNA clones encoding the 90 kD antigen, JENDOUBI and BONNEFOY (1988) used a polyclonal serum raised against gel purified parasite proteins of ~ 90 kD to screen a pUC9 expression library. They isolated several clones reactive with the antiserum and determined the sequence of one clone, C27. The deduced amino acid sequence of the open reading frame in clone C27 was 150 amino acids in length and displayed 50%–56% identity with members of the hsp90 family. The genomic organization and regulation of expression of this gene were not described. It must also be noted here that it has not yet been shown that the hsp90 homologue encoded by clone C27 actually reacts with antibodies from infected animals or is involved in a protective immune response.

3 The hsp70 Family

Perhaps the most highly conserved and abundant of the heat shock proteins is the hsp70 family (reviewed in LINDQUIST 1986; LINDQUIST and CRAIG 1988). The amino acid sequence of the human hsp70 has 50% identity with that of the *E. coli* hsp70 and 73% identity with that of the *Drosophila* hsp70. Interestingly, sequence identity is highest in the amino terminal region of the protein, which has been suggested to have a conserved ATPase activity. The *E. coli* homologue is encoded by a single gene, *dnaK*. However, eukaryotic cells contain a number of hsp70 genes and proteins. Some are constitutively expressed (heat shock cognate proteins or hsc70), some are heat inducible (hsp70), and some are glucose regulated (grp78). These proteins can be found in the nucleus, cytoplasm, and endoplasmic reticulum. Current models for the functions of hsp70 proteins are that they interact with other proteins in an ATP-dependent manner to promote folding or unfolding or to facilitate transport across membranes (LANGER and NEUPERT, this volume). Members of the hsp70 family have been shown to be antigens of several parasites and bacteria.

hsp70 Antigens in Parasistes. In squirrel monkeys, antibodies to 72-kD proteins of *P. falciparum* are correlated with immune protection (DUBOIS et al. 1984). The 72-kD proteins have been identified as members of the hsp70 family. The genes or portions of the genes for at least four members of the *P. falciparum* hsp70 family have been cloned and sequenced (BIANCO et al. 1986; ARDESHIR et al. 1987; KUMAR et al. 1988; MATTEI et al. 1988; PETERSON et al. 1988; KUN and MULLER-HILL 1989). The proteins encoded by these genes display 50%–70% identity with each other and with hsp70 proteins from other species. Where studied, the cloned genes appear to be constitutively expressed and are found in all blood stages of the malaria parasite. One of the proteins, Ag63, contains a region of tandemly repeated amino acid sequence. Similar, although shorter, repeats are seen in hsp70 proteins from *Plasmodium chabaudi* (SHEPPARD et al. 1989), *Brugia pahangi* (SELKIRK et al. 1989), and the rat (SORGER and PELHAM 1987), but not in another hsp70 from *P. falciparum*, Ag361 (PETERSON et al. 1988). However, unlike other malaria antigens, the repeated regions of Ag63 do not appear to be immunogenic in humans (PETERSON et al. 1988). Furthermore, although the two *P. falciparum* proteins Ag63 and Ag361 display impressive sequence identity, antibodies from pooled human sera affinity purified using either protein do not cross-react with the other protein (PETERSON et al. 1988). This suggests that the antibody response to the malarial hsp70 proteins is directed mainly towards nonconserved epitopes. However, human and monkey immune sera do contain reactivities to sequences that are shared between *P. falciparum* and a human hsp70 (MATTEI et al. 1989; RICHMAN et al. 1989). Thus, at least some conserved epitopes are immunogenic during a naturally occurring infection.

Virtually all sera from humans and animals infected with *S. mansoni* show reactivity with a 70-kD antigen (HEDSTROM et al. 1987, 1988). In mice, antibodies to this protein arise shortly after infection and are correlated with immune protection (HEDSTROM et al. 1987). Using pooled sera from infected humans, HEDSTROM et al. (1987) isolated recombinant DNA clones that expressed this 70-kD protein. The deduced amino acid sequence has impressive homology with human (80%) and *Drosophila* (71%) hsp70 proteins. There are four copies of the *hsp70* genes in schistosomes (SCALLON et al. 1987). The cloned *hsp70* genes appear to be constitutively expressed in larval and adult schistosomes, although higher levels of expression can be found in some cell types and under certain conditions, such as in transforming schistosomulae (reviewed in NEWPORT et al. 1988a). At least one *hsp70* gene of *S. mansoni* is heat inducible (JOHNSON et al. 1989). The immune response to the 70-kD antigen appears to be directed mainly towards nonconserved sequences (SCALLON et al. 1987; HEDSTROM et al. 1988). That is, antibodies in the sera of persons infected with *S. japonicum* do not react with the *S. mansoni* hsp70 homologue, although they do react strongly with the *S. japonicum* hsp70 homologue. Because of the widespread and apparently specific antibody response to this antigen, it has been suggested that this antigen may be useful in immunodiagnosis (NEWPORT et al. 1988b).

Most individuals infected with *Brugia malayi* (Brugian filariasis) produce antibodies reactive with a 70-kD antigen (Selkirk et al. 1989). The gene encoding this protein has been isolated from a *B. pahangi* cDNA library and shown to be part of a multigene family (SELKIRK et al. 1989). The cloned protein is homologous to members of the hsp70 family, particularly a rat hsc70 (85% amino acid sequence identity). The *B. pahangi hsp70* gene is expressed in all stages of the parasite although its expression is heat inducible in the larval stage. The majority of the immune response to this protein in infected persons seems to be directed against epitopes found specifically on the *Brugia* protein. However, sera from some infected persons do cross-react with hsp70 molecules from *P. falciparum* and. *S. mansoni* (SELKIRK et al. 1989).

The hsp70 proteins of *Wuchereria bancrofti*, another causative agent of lymphatic filariasis, also appear to be immunogenic. The evidence for this is that sera from persons infected with *W. bancrofti* react with hsp70 homologues from *B. pahangi* and *Onchocerca volvulus*, the causative agent of ocular filariasis (SELKIRK et al. 1989; ROTHSTEIN et al. 1989). Indeed, ROTHSTEIN et al. (1989) used sera from *W. bancrofti* infected persons to isolate a cDNA clone of *O. volvulus* that expressed an antigen reactive with sera from amicrofilaremic individuals. The deduced amino acid sequence of the *O. volvulus* protein revealed that it is a member of the hsp70 family. However, neither the immunoreactivity of the *O. volvulus* protein during an *O. volvulus* infection nor of the *W. bancrofti* protein during a *W. bancrofti* infection has been reported. Nonetheless, these results indicate that conserved epitopes of hsp70 can be immunoreactive.

A 70-kD surface protein of *T. cruzi* reacts strongly with sera from chronically infected mice. When a rabbit antiserum raised against purified surface antigen was used to screen a cDNA library, eight clones were isolated that encoded part of a 71-kD polypeptide (ENGMAN et al. 1989). The deduced amino acid sequence of the cloned protein has 55% identity with the *E. coli* hsp70 and 45% identity with eukaryotic hsp70s. However, the cloned hsp70 homologue probably is not the 70-kD surface antigen—the surface antigen does not react with antibodies elicited by the cloned protein. Apparently the rabbit antiserum contained antibodies reactive with hsp70 either because the hsp70 homologue copurified with the surface protein or because they have cross-reactive epitopes. Thus, the antigenicity of the hsp70 homologue during a *T. cruzi* infection is uncertain.

Members of the hsp70 family of a related parasite protozoan, *Leishmania major*, have also been suggested to be antigens (SEARLE et al. 1989; D. YOUNG et al. 1989). However, details of this antigenicity have not yet been published.

hsp70 Antigens in Bacteria. An immunoreactive 70-kD antigen has been identified in *Mycobacterium leprae, M. tuberculosis,* and *M. bovis* BCG (reviewed in BRITTON et al. 1986a; WATSON 1989). These 70-kD proteins react strongly with antibodies in sera from persons with leprosy or tuberculosis as well as from experimental animals. These proteins are potent T cell antigens that can elicit a delayed type hypersensitivity response in infected persons.and can stimulate peripheral blood lymphocytes to proliferate and produce interferon-γ (BRITTON et al. 1986b). The genes encoding the 70-kD antigens have been cloned from *M. leprae, M. tuberculosis,* and *M. bovis* BCG and their sequences determined (R. YOUNG et al. 1985a, b; SHINNICK et al. 1987a; D. YOUNG et al. 1988a; GARSIA et al. 1989). The mycobacterial proteins are highly homologous to each other and to members of the hsp70 family (D. YOUNG et al. 1988a; GARSIA et al. 1989). The expression of the *M. bovis* BCG hsp70 homologue is heat inducible (MEHLERT and YOUNG 1989). Both conserved and nonconserved epitopes are immunoreactive since T cells and antibodies have been isolated from humans that cross-react with hsp70s from *E. coli* and humans as well as ones that react specifically with the mycobacterial proteins (BRITTON et al. 1986b; REES et al. 1988; TSOUFLA et al. 1989).

A 75-kD protein of *Chlamydia trachomatis* is recognized by antibodies in sera of over 50% of infected persons. Serum antibody levels to this protein have been correlated with protection against ascending fallopian tube infection (BRUNHAM et al. 1987). This protein also appears to be surface accessible, and monospecific antibodies directed against it can neutralize infectivity (MACLEAN et al. 1988; DANILITION et al. 1990). The protein has amino acid sequence homology with the *E. coli* DnaK protein and eukaryotic hsp70s (DANILITION et al. 1990). Studies with patient sera and rabbit sera suggest that the immune response to this protein is directed primarily at nonconserved epitopes (BIRKELUND et al. 1989).

4 The hsp60 or GroEL Family

The hsp60 proteins have been extensively characterized in bacteria and only recently have their eukaryotic homologues been identified (JINDAL et al. 1989; MIZZEN et al. 1989). The hsp60 amino acid sequences are highly conserved between bacteria and man, with 50%–60% sequence identity. It has been suggested that hsp60 proteins play roles in promoting protein folding, assembly of oligomeric protein structures, and transport of proteins across membranes (HEMMINGSEN et al. 1988; WELCH et al. 1989). These proteins are antigens of any pathogenic bacteria, but as yet no parasite antigens have been identified as members of the hsp60 family. The hsp60 proteins do not appear to elicit a protective immune response.

Two sets of studies on different groups of bacteria led to the realization that a prominent bacterial antigen was actually a heat shock protein. First, independent studies by KAIJSER (1975) and HOIBY (1975) identified an ~60-kD protein as a widely cross-reactive antigen of *E. coli* and of *Pseudomonas aeruginosa*. This protein was named "common antigen." Subsequent studies showed that proteins cross-reactive with common antigen are present in virtually all bacteria. Second, it was realized that the highly immunoreactive 65-kD proteins of *M. tuberculosis* and *M. bovis* BCG are the mycobacterial counterparts to common antigen and that these antigens are homologous to the *E. coli* GroEL heat shock protein (SHINNICK et al. 1988; THOLE et al. 1988; D. YOUNG et al. 1988a). Finally, it has been shown that the genes for these antigens are heat inducible in *E. coli* (LINDQUIST 1986), *P. aeruginosa* (ALLAN et al. 1988), and *M. smegmatis* (SHINNICK et al. 1988).

Besides the studies on common antigen, much of the evidence that hsp60 is a major (immunodominant?) antigen comes from studies on mycobacteria (reviewed in D. YOUNG et al. 1987, 1988b; R. YOUNG 1990). The evidence includes that (a) antibodies and T cells that react with hsp60 are easily isolated from sera of persons infected with *M. tuberculosis* or *M. leprae* and (b) monoclonal antibodies to hsp60 are frequently isolated from mice immunized with whole mycobacteria or crude sonicates. An additional key observation was that 20% of the mycobacteria-reactive T cells in the *M. tuberculosis* immunized mouse reacted with hsp60 (KAUFMANN et al. 1987).

The humoral immune response to hsp60 is directed against both conserved and nonconserved epitopes. For example, monoclonal antibodies have identified epitopes that are specific for the hsp60 of *M. leprae* or *M. tuberculosis* as well as epitopes that are shared with hsp60 homologues from bacteria and humans (MEHRA et al. 1986; SHINNICK et al. 1988; THOLE et al. 1988; ANDERSON et al. 1988; DUDANI and GUPTA 1989). Along similar lines, PLIKAYTIS et al. (1987) showed that an antiserum to *Legionella* hsp60 contained antibodies that reacted with both types of epitopes by removing the cross-reactive antibodies by sequential adsorption with *P. aeruginosa*, *P. fluorescens*, and *Bordetella pertussis* to generate an antiserum that reacted specifically with *Legionella* hsp60. The cellular immune response to hsp60 is also directed against conserved and nonconserved epitopes. Of particular interest here is the observation that T cells can be isolated from tuberculosis patients as well as from apparently uninfected individuals that cross-react with hsp60s from *E. coli*, mycobacteria, and humans (MUNK et al. 1988, 1989; LAMB et al. 1989).

Immunoreactive hsp60 homologues have been found in species of *Borrelia* (HANSEN et al. 1988), *Chlamydia* (MORRISON et al. 1989), *Coxiella* (VODKIN and WILLIAMS 1988), *Legionella* (BANGSBORG et al. 1989; SAMPSON et al. 1990), *Neisseria* (PANNEKOEK et al. 1988), *Salmonella* (BUCHMEIER and HEFFRON 1990), *Rickettsia* (STOVER et al. 1990), and *Treponema* (HINDERSSON et al. 1987).

5 Low Molecular Weight Heat Shock Proteins

Most organisms produce one or more low molecular weight heat shock proteins ranging in size from 14 to 30 kD (reviewed in LINDQUIST 1986; LINDQUIST and CRAIG 1988; WELCH et al. 1989). In *Drosophila*, the members of this family include hsp22, hsp23, hsp26, and hsp28. There is a single major member of this family in yeast, hsp26. The eukaryotic small heat shock proteins are in general encoded by multigene families and exist in several isoforms, some of which are phosphorylated. The different isoforms and genes are expressed at different times during development and in different cell types. The small heat shock proteins display homology with α-crystallin proteins and have similar hydropathy profiles. The similarities with α-crystallin proteins may relate to their ability to form complexes with other proteins and RNA. The function of these proteins is unclear. Surprisingly, a yeast mutant deleted for the hsp26 gene is phenotypically indistinguishable from the wild-type strain except for the absence of hsp26 (SUSEK and LINDQUIST 1989). One parasite and one bacterial antigen have been identified as members of this family of heat shock proteins.

A prominent 40-kD antigen of the egg stage of S. mansoni can be immunoprecipitated by antibodies from sera of > 90% of infected individuals (TAYLOR et al. 1984; NENE et al. 1986). This protein is probably encoded by a multigene family, and its expression is differentially regulated during the parasite life cycle (NENE et al. 1986). The amino acid sequence of this protein has two regions that are homologous to small heat shock and α-crystallin proteins, and which are separated by a region that is homologous to the Drosophila hsp22 (DE JONG et al. 1988). The authors suggest that these sequence features reflect an ancient intragenic tandem duplication. The immune response to this antigen appears to be directed against specific epitopes and as such may be useful in immunodiagnosis (CORDINGLEY et al. 1984; NEWPORT et al. 1988a).

An 18-kD antigen of M. leprae appears to be a particularly good T cell immunogen and antigen in a large proportion of individuals immunized with a killed M. leprae vaccine and in 93% of long-term leprosy contacts, i.e., individuals presumably immune to M. leprae (DOCKRELL et al. 1989). The protein contains T cell and antibody binding sites that are unique to M. leprae as well as ones that are shared with other Mycobacterium species (DOCKRELL et al. 1988; HARRIS et al. 1989). The amino acid sequence of the 18-kD antigen has homology with members of the low molecular weight heat shock family, including soybean hsp17.5E, Drosophila hsp22, and the schistosome p40 protein (BOOTH et al. 1988; NERLAND et al. 1988).

6 The GroES Family

In many bacteria, including E. coli, hsp60 or GroEL is expressed as part of a heat shock operon that also contains a 10-kD protein called GroES (HEMMINGSEN et al. 1988). In eukaryotes, hsp60 homologues are not part of operons and GroES homologues have not been described (HEMMINGSEN et al. 1988; JINDAL et al. 1989). In mycobacteria, the GroES homologue is antigenic.

An ~ 10 kD protein of M. bovis BCG reacts with antibodies and T cells elicited by M. tuberculosis and M. bovis (MINDEN et al. 1984). A similarly sized protein of

M. leprae is also antigenic (V. MEHRA, personal communication). The amino acid sequences of the 10-kD proteins have ~ 45 % identity with the sequence of the *E. coli* GroES protein (BAIRD et al. 1988; SHINNICK et al. 1989). Incidentally, the genes for these two proteins are in separate operons in mycobacteria (MEHRA et al. 1986; SHINNICK 1987; THOLE et al. 1987; SHINNICK et al. 1989). It should be noted here that the *Coxiella burnetti* GroES homologue does not appear to be antigenic (VODKIN and WILLIAMS 1988).

7 Why are Heat Shock Proteins Immunoreactive?

The immunoreactivity of heat shock proteins raises questions as to why such highly conserved proteins should be the targets of an immune response. Possible explanations for this include that heat shock proteins (a) are abundant cellular proteins, (b) have conserved epitopes that may prime the host for an immune response to these proteins, (c) may be preferentially processed for presentation due to either structural or functional features, and (d) may be virulence factors. In the discussion of each of these points below, it should be noted that the hypotheses are speculative and at present have little or no experimental support, although several are readily testable. The reader is referred to reviews by PALLEN (1989), D. YOUNG et al. (1989), YOUNG and ELLIOTT (1989), and R. YOUNG (1990) for additional discussions of the immunoreactivity of heat shock proteins. *Abundance.* During the growth of *E. coli* at 37 °C, hsp60 accounts for ~ 1.6 % and hsp70 for ~ 1.4 % of total cell protein (NEIDHARDT et al. 1984). These proteins accumulate to even higher levels in bacteria undergoing stress; hsp70, hsp60, and GroES can account for more than 15 % of cell protein in heat shocked cells (NEIDHARDT et al. 1984). In mycobacteria, hsp60 accounts for ~ 9 % of cell protein when bacilli are grown in zinc deficient media (DE BRUYN et al. 1987). Although hsp70 and hsp60 do not accumulate to such high levels in eukaryotic cells, they are still abundant proteins. For example, in adult schistosomes, hsp70 accounts for ~ 1 % of cell protein and is even more abundant in transforming schistosomulae (NEWPORT et al. 1988a).

The synthesis and accumulation of subsets of the heat shock proteins can also be stimulated by anoxia, amino acid or glucose deprivation, and exposure to hydrogen peroxide, ethanol, or heavy metal ions (reviewed in NEIDHARDT et al. 1984; LINDQUIST 1986; MORGAN et al. 1986; CRAIG and LINDQUIST 1988). (This is why "stress response" proteins is a more appropriate designation than heat shock proteins.) Thus, not only may the change in temperature that occurs upon infection induce protein synthesis but other factors may also stimulate production and accumulation of heat shock proteins. Indeed, BUCHMEIER and HEFFRON (1990) have shown that even in the absence of a heat shock the two most abundantly expressed proteins of *Salmonella typhimurium* following phagocytosis are hsp70 and hsp60. Thus, the stress that is inherent inside the lysosome of a macrophage might help to induce the synthesis of these proteins to high levels.

Overall then, the immunoreactivity of heat shock proteins could be due simply to their being major constituents of a pathogen when that pathogen is within an antigen presenting cell and hence the major constituents that are processed and presented to the immune system.

Immunological Priming. Since heat shock proteins are highly conserved, infection by one pathogen might prime the host for an immune response against these proteins upon infection with a second pathogen. This suggests that the immune response should be directed primarily against conserved epitopes. However, this does not seem to be the case for antibody responses. Although antibodies directed against both conserved and nonconserved epitopes of hsp60 are found following bacterial infections, the antibody response to parasite hsp70 seems to be predominantly directed against nonconserved epitopes. On the other hand, T cells that recognize either conserved or nonconserved epitopes of the mycobacterial hsp60 or hsp70 homologues can be readily isolated (reviewed in R. YOUNG 1990).

The existence of T cells reactive with conserved epitopes raises the possibility that a host might be primed for a helper T cell response to conserved epitopes. If so, such primed helper T cells may be able to promote a humoral immune response to an associated conserved or nonconserved B cell epitope. In this regard, it is interesting to note that covalent attachment of a strongly immunogenic T cell epitope of the *M. tuberculosis* hsp60 to a poorly immuno-genic epitope of the foot-and-mouth disease virus greatly increased the antibody response to the virus epitope (COX et al. 1988).

Structural and Functional Considerations. Although most heat shock proteins appear to be located intracellularly, some members of the hsp90, hsp70, and hsp60 families appear to be surface accessible. Also, certain cells undergoing heat shock or other stresses can express heat shock proteins, including hsp60, on their surfaces or even secrete them (KOGA et al. 1989; JARJOUR et al. 1989; DE BRUYN et al. 1987). In addition, heat shock proteins can associate with other cellular proteins and can form large oligomeric structures (reviewed in LINDQUIST and CRAIG 1988; WELCH et al. 1989). For example, hsp60 forms a homomultimer of > 300 kD and hsp70 interacts with DNA replication complexes and clathrin. Perhaps, because of these structural features, heat shock proteins can be efficiently processed and presented on the macrophage surface, and hence, become readily available for interaction with other components of the immune system.

Two of the postulated functions of heat shock proteins may also facilitate their processing and presentation. That is, members of the hsp70 family have been suggested to play roles in targeting intracellular proteins for lyso-somal degradation (CHIANG et al. 1989) and in antigen presentation on the macrophage cell surface (VANBUSKIRK et al. 1989). Perhaps the conserved domains of the pathogen's heat shock proteins will target them to the organelles and proteins involved in antigen processing and presentation and thereby facilitate their processing. Presentation might also be facilitated by the ability of hsp70 and hsp60 to bind to denatured proteins. Perhaps, a pathogen's heat shock proteins could be "coprocessed" along with other proteins and thereby increase the likelihood of a heat shock protein being presented by the macrophage.

Overall then, heat shock proteins have structural and functional features that may make them particularly good candidates for processing and presentation to the immune system, and hence, potential immunogens.

Virulence Determinants. Exposure of *Leishmania braziliensis* promastigotes to a heat shock increases their infectivity and pathogenicity (SMEJKAL et al. 1988). Also, the expression of stress response proteins may be one key to the ability of intracellular pathogens such as *M. tuberculosis* or *S. typhimurium* to survive within cells. Indeed, mutants of *S. typhimurium* that do not increase the expression of stress response proteins upon phagocytosis are less able to survive within macrophages and less virulent in vivo than the wild-type strain (BUCHMEIER and HEFFRON 1990). If heat shock proteins are "virulence factors," then it might be advantageous for the host to target these proteins for an immune response. A discussion of this aspect of heat shock proteins is presented in the chapter by D. YOUNG.

8 Concluding Remarks

It is clear that heat shock proteins are immunologically active components of many pathogens. The list of antigenic heat shock proteins is probably larger than that presented in Table 1. Indeed, if one reads the literature with an eye towards finding antigenic heat shock proteins, one can identify antigens in many pathogens that share features with heat shock proteins and vice versa, but the experimental evidence to link the two is not available. However, one must keep in mind that simply cloning a heat shock protein using antibodies to bacterial or parasitic antigens is not sufficient to identify the heat shock protein as being immunoreactive during the course of a naturally occurring infection (e.g., see ENGMAN et al. 1989).

For several pathogens, the humoral immune response to the heat shock proteins is directed predominantly towards nonconserved epitopes. Perhaps, as suggested by NEWPORT et al. (1988a) and others, this is a parasite survival strategy. Here, the nonconserved epitopes might serve as an immunological smokescreen to divert the host's immune response away from conserved epitopes, which may be regions required for functional activity. Of course, this is not a general survival strategy since some heat shock proteins in some pathogens seem to be targets of a protective immune response. On the other hand, several investigators have taken advantage of this feature of the humoral immune response to develop immunodiagnostic tests. Although assays

Table 1. Antigenic heat shock homologues

Family	Species	Size (kD)
hsp90 family	Plasmodium falciparum	90
	Schistosoma mansoni	86
	Trypanosoma cruzi	85
hsp70 family	Plasmodium falciparum	70–75
	Schistosoma mansoni	70
	Schistosoma japonicum	70
	Onchocerca volvulus	70
	Brugia pahangi	70
	Leishmania major	70
	Chlamydia trachomatis	75
	Mycobacterium leprae	70
	Mycobacterium tuberculosis	71
hsp60 family	Mycobacterium leprae	65
	Mycobacterium tuberculosis	65
	Borrelia burgdorferi	60
	Chlamydia trachomatis	57
	Coxiella burnetii	60
	Legionella pneumophila	60
	Pseudomonas aeruginosa	60
	Rickettsia tsutsugamushi	58
	Treponema pallidum	60
Low molecular	Schistosoma mansoni	40
weight hsp's	Mycobacterium leprae	18
GroES family	Mycobacterium leprae	12
	Mycobacterium tuberculosis	12

involving the whole protein are usually too cross-reactive to be diagnostic, epitope-specific assays may yet turn out to be useful.

Finally, heat shock proteins of pathogenic microorganisms can elicit humoral and cellular immune response to epitopes that are shared with their hosts. Such immunoreactivity could play important roles in pathogenicity and autoimmune consequences of infections. These features of the immune response to heat shock proteins are discussed in detail elsewhere in this volume.

References

Allan B, Linseman M, MacDonald LA, Lam JS, Kropinski AM (1988) Heat shock response in *Pseudomonas aeruginosa*. J Bacteriol 170: 3668–3674

Anderson DC, Barry MG, Buchanan TM (1988) Exact definition of species-specific and cross-reactive epitopes of the 65 kDa protein of *Mycobacterium leprae* using synthetic peptides. J Immunol 141: 607–613

Ardeshir F, Flint JE, Richman J, Reese RT (1987) A 75-kD merozoite surface protein of *Plasmodium falciparum* which is related to the 70-kD heat shock proteins. EMBO J 6: 493–499

Baird PN, Hall LM, Coates AR (1988) A major antigen from *Mycobacterium tuberculosis* which is homologous to the heat shock proteins GroES from *E. coli* and the *hptA* gene product of *Coxiella burnetti*. Nucleic Acids Res 16: 9047

Bangsborg JM, Collins MT, Hoiby N, Hindersson P (1989) Cloning and expression of the *Legionella micdadei* "common antigen" in *Escherichia coli*. Acta Pathol Microbiol Immunol Scand [ ] 97: 14–22

Bianco AE, Favaloro JM, Burkot TR, Culvenor JG, Crewther PE, Brown GV, Anders RF, Coppel RL, Kemp DJ (1986) A repetitive antigen of *Plasmodium falciparum* that is homologous to heat shock protein 70 of *Drosophila melanogaster*. Proc Natl Acad Sci USA 83: 8713–8717

Birkelund S, Lundemose AG, Christiansen G (1989) Characterization of native and recombinant 75-kilodalton immunogens from *Chlamydia trachomatis* serovar L2. Infect Immun 57: 2683–2690

Booth RJ, Harris DP, Love JM, Watson JD (1988) Antigenic proteins of *Mycobacterium leprae*: complete sequence of the 18 kDa protein. J Immunol 140: 597–601

Britton WJ, Garsia RJ, Hellqvist L, Watson JD, Basten A (1986a) The characterization and immunoreactivity of a 70 kD protein common to *Mycobacterium leprae* and *Mycobacterium bovis* BCG. Lepr Rev 57 [Suppl 2]: 67–75

Britton WJ, Hellqvist L, Basten A, Inglis AS (1986b) Immunoreactivity of a 70-kD protein purified from *Mycobacterium bovis* bacillus Calmette-Guerin by monoclonal antibody affinity chromatography. J Exp Med 164: 695–708

Brunham RC, Peeling R, Maclean I, McDowell J, Persson K, Osser S (1987) Postabortal *Chlamydia trachomatis* salpingitis: correlating risk with antigen-specific serological responses and with neutralization. J Infect Dis 155: 749–755

Buchmeier NA, Heffron F (1990) *Salmonella* proteins induced following phagocytosis by macrophages are controlled by multiple regulons. Science 248: 730–732

Cheng M, Hartl FU, Martin J, Pollock R, Kalousek F, Neupert W, Hallberg E, Hallberg R, Horwick A (1989) Mitochondrial heat-shock protein hsp60 is essential for assembly of proteins imported into yeast mitochondria. Nature 337: 620–625

Chiang HL, Terlecky SR, Plant CP, Dice JF (1989) A role for a 70-kilodalton heat shock protein in lysosomal degradation of intracellular proteins. Science 246: 382–385

Cordingley JS, Taylor DW, Dunne DW, Butterworth AE (1984) Clone banks of cDNA from the parasite *Schistoma mansoni*: isolation of clones containing a potentially immunodiagnostic antigen gene. Gene 26: 25–39

Cox JH, Ivanyi J, Young DB, Lamb JR, Syred AD, Francis MJ (1988) Orientation of epitopes influences the immunogenicity of synthetic peptide dimers. Eur J Immunol 18: 2015–2019

Danilition SL, Maclean IW, Peeling R, Winston S, Brunham RC (1990) The 75-kilodalton protein of *Chlamydia trachomatis*: a member of the heat shock protein 70 family? Infect Immun 58: 189–196

de Bruyn J, Bosmans R, Turneer M, Weckx M, Myabenda J, Van Vooren JP, Falmagne P, Wiker HG, Harboe M (1987) Purification, partial characterization, and identification of a skin reactive protein antigen of *Mycobacterium bovis* BCG. Infect Immun 55: 245–252

de Jong WW, Leunissen JAM, Leenen PJM, Zweers A, Versteeg M (1988) Dogfish α-crystallin sequences. Comparison with small heat shock proteins and *Schistosoma* egg antigen. J Biol Chem 263: 5141–5149

Dockrell HM, Stoker NG, Lee SP, Jackson M, Grant KA, Jouy NF, Lucas SB, Hasan R, Hussain R, McAdam KPWJ (1989) T-cell recognition of the 18-kilodalton antigen of *Mycobacterium leprae*. Infect Immun 57: 1979–1983

Dragon EA, Sias SR, Kato EA, Gabe JD (1987) The genome of *Trypanosoma cruzi* contains a constitutively expressed, tandemly arranged multicopy gene homologous to a major heat shock protein. Mol Cell Biol 7: 1271–1275

Dubois P (1989) Heat shock proteins and immunity. Res Immunol 140: 653–659

Dubois P, Dedet JP, Fandeur T, Roussilhon C, Jendoubi M, Pauillac S, Mercereau-Puijalon O, Pereira de Silva L (1984) Protective immunization of the squirrel monkey against asexual blood stages of *Plasmodium falciparum* by use of parasite protein fractions. Proc Natl Acad Sci USA 81: 229–233

Dudani AK, Gupta RS (1989) Immunological characterization of a human homolog of the 65-kilodalton mycobacterial antigen. Infect Immun 57: 2786–2793

Engman DM, Kirchhoff LV, Donelson JE (1989) Molecular cloning of mtp70, a mitochondrial member of the hsp70 family. Mol Cell Biol 9: 5163–5168

Garsia RJ, Hellqvist L, Booth RJ, Radford AJ, Britton WJ, Astbury L, Trent RJ, Basten A (1989) Homology of the 70-kilodalton antigens of *Mycobacterium leprae* and *Mycobacterium bovis* with the *Mycobacterium tuberculosis* 71-kilodalton antigen and with the conserved heat shock protein 70 of eukaryotes. Infect Immun 57: 204–212

Hansen K, Bangsborg JM, Fjordvang H, Pedersen NS, Hindersson P (1988) Immunochemical characterization and isolation of the gene for a *Borrelia burgdorferi* immunodominant 60 kilodalton antigen common to a wide range of bacteria. Infect Immun 56: 2047–2053

Harris DP, Backstrom BT, Booth RJ, Love SG, Harding DR, Watson JD (1989) The mapping of epitopes of the 18-kDa protein of *Mycobacterium leprae* recognized by murine T cells in a proliferation assay. J Immunol 143: 2006–2012

Hedstrom R, Culpepper J, Harrison RA, Agabian N, Newport G (1987) A major immunogen in *Schistosoma mansoni* infections is homologous to the heat-shock protein Hsp70. J Exp Med 165: 1430–1435

Hedstrom R, Culpepper J, Schinski V, Agabian N, Newport G (1988) Schistosome heat-shock proteins are immunologically distinct host-like antigens. Mol Biochem Parasitol 29: 275–282

Hemmingsen SM, Woolford C, van der Vies SM, Tilly K, Dennis DT, Georgopoulos CP, Hendrix RW, Ellis RJ (1988) Homologous plant and bacterial proteins chaperone oligomeric protein assembly. Nature 333: 330–334

Hindersson P, Petersen CS, Petersen NS, Hoiby N, Axelsen N (1984) Immunologic cross-reaction between antigen Tp-4 of *Treponema pallidum* and an antigen common to a wide range of bacteria. Acta Pathol Microbiol Immunol Scand [B] 92: 183–188

Hindersson P, Knudsen JD, Axelsen N (1987) Cloning and expression of *Treponema pallidum* common antigen (Tp-4) in *Escherichia coli* K12. J Gen Microbiol 133: 587–596

Hoiby N (1975) Cross-reactions between *Pseudomonas aeruginosa* and thirty-six other bacterial species. Scand J Immunol 4 [Suppl 2]: 187–196

Jarjour W, Tsai V, Woods V, Welch W, Pierce W, Shaw M, Mehta H, Dillman W, Zvaifler N, Winfield J (1989) Cell surface expression of heat shock proteins. Arthritis Rheum 32: S44

Jendoubi M, Bonnefoy S (1988) Identification of a heat shock-like antigen in *P. falciparum*, related to the heat shock protein 90 family. Nucleic Acids Res 16: 10928

Jendoubi M, Dubois P, Pereira de Silva L (1985) Characterization of one polypeptide antigen potentially related to protective immunity against the blood infection by *Plasmodium falciparum* in the squirrel monkey. J Immunol 134: 1941–1945

Jindal S, Dudani AK, Singh B, Harley CB, Gupta RS (1989) Primary structure of a human mitochondrial protein homologous to the bacterial and plant chaperonins and to the 65-kilodalton mycobacterial antigen. Mol Cell Biol 9: 2279–2283

Johnson KS, Wells K, Bock JV, Nene V, Taylor DW, Cordingley JS (1989) The 86-kilodalton antigen from *Schistosoma mansoni* is a heat-shock protein homologous to yeast hsp-90. Mol Biochem Parasitol 36: 19–28

Kaijser B (1975) Immunologic studies of an antigen common to many gram negative bacteria with special reference to *E. coli*. Int Arch Allergy Appl Immunol 48: 72–81

Kaufmann SHE, Vath U, Thole JER, van Embden JDA, Emmrich F (1987) Enumeration of T cells reactive with *M. tuberculosis* organisms and specific for the recombinant mycobacterial 64 kilodalton protein. Eur J Immunol 17: 351–357

Koga T, Wand-Wurttenberger A, de Bruyn J, Munk ME, Schoel B, Kaufmann SHE (1989) T cells against a bacterial heat shock protein recognize stressed macrophages. Science 245: 1112–1115

Kumar N, Syin C, Carter R, Quakyi I, Miller LH (1988) *Plasmodium falciparum* gene encoding a protein similar to the 78-kDa rat glucose-regulated stress protein. Proc Natl Acad Sci USA 85: 6277–6281

Kun J, Muller-Hill B (1989) The sequence of a third member of the heat shock protein family in *Plasmodium falciparum*. Nucleic Acids Res 17: 5384

Lamb JR, Bal V, Mendez-Sampirio P, Mehlert A, So A, Rothbard J, Jindal S, Young RA, Young DB (1989) Stress proteins may provide a link between the immune response to infection and autoimmunity. Int Immunol 1: 191–196

Lindquist S (1986) The heat-shock response. Ann Rev Biochem 55: 1151–1191

Lindquist S, Craig EA (1988) The heat-shock proteins. Ann Rev Genet 22: 631–637

Maclean IW, Peeling RW, Brunham RC (1988) Characterization of *Chlamydia trachomatis* antigens with monoclonal and polyclonal antibodies. Can J Microbiol 34: 141–147

Mattei D, Osaki LS, Pereira da Silva L (1988) A *Plasmodium falciparum* gene encoding a heat shock-like antigen related to the rat 78 kD glucose-regulated protein. Nucleic Acids Res 16: 5204

Mattei D, Scherf A, Bensaude O, Pereira da Silva L (1989) A heat shock-like protein from the human malaria parasite *Plasmodium falciparum* induces autoantibodies. Eur J Immunol 19: 1823–1828

Mehlert A, Young DB (1989) Biochemical and antigenic characterization of the *Mycobacterium tuberculosis* 71 kD antigen, a member of the 70 kD heat-shock protein family. Mol Microbiol 3: 125–130

Mehra V, Sweetser D, Young RA (1986) Efficient mapping of protein antigenic determinants. Proc Natl Acad Sci USA 83: 7013–7017

Minden P, Kelleher PJ, Freed J, Nielsen LD, Brennan PJ, McPheron L, McClatchy JK (1984)

Immunological evaluation of a component isolated from *Mycobacterium bovis* BCG with a monoclonal antibody to *M. bovis* BCG. Infect Immun 46: 519–525

Mizzen LA, Chang C, Garrels JI, Welch WJ (1989) Identification, characterization, and purification of two mammalian stress proteins present in mitochondria, grp75, a member of the hsp70 family and hsp58, a homologue of the bacterial GroEL protein. J Biol Chem 264: 20664–20675

Morgan RW, Christman MF, Jacobson FS, Storz G, Ames BN (1986) Hydrogen peroxide-inducible proteins in *Salmonella typhimurium* overlap with heat shock and other stress proteins. Proc Natl Acad Sci USA 83: 8059–8063

Morrison RP, Belland RJ, Lying K, Caldwell HD (1989) Chlamydial disease pathogenesis. The 57-kD chlamydial hypersensitivity antigen is a stress response protein. J Exp Med 170: 1271–1283

Munk ME, Schoel B, Kaufmann SHE (1988) T-cell responses of normal individuals towards recombinant protein antigens of *Mycobacterium tuberculosis*. Eur J Immunol 18: 1835–1838

Munk ME, Schoel B, Modrow S, Karr RW, Young RA, Kaufmann SHE (1989) T lymphocytes from healthy individuals with specificity to self epitopes shared by the mycobacterial and human 65-kilodalton heat shock protein. J Immunol 143: 2844–2849

Neidhardt FC, VanBogelen RA, Vaughn V (1984) The genetics and regulation of heat-shock proteins. Ann Rev Genet 18: 295–329

Nene V, Dunne DW, Johnson KS, Taylor DW, Cordingley JS (1986) Sequence and expression of a major egg antigen from *Schistosoma mansoni*. Homologies to heat shock proteins and alpha-crystallins. Mol Biochem Parasit 21: 179–188

Nerland AH, Mustafa AS, Sweetser D, Godal T, Young RA (1988) A protein antigen of *Mycobacterium leprae* is related to a family of small heat shock proteins. J Bacteriol 170: 5919–5921

Newport G, Culpepper J, Agabian N (1988a) Parasite heat shock proteins. Parasitol Today 4: 306–312

Newport GR, Hedstrom RC, Kallestad J, Tarr P, Klebanoff S, Agabian N (1988b) Identification, molecular cloning, and expression of a schistosome antigen displaying diagnostic potential. Am J Trop Med Hyg 38: 540–546

Pallen M (1989) Bacterial heat-shock proteins and serodiagnosis. Serodiag Immunother Infect Dis 3: 149–159

Pannekoek Y, Weel JFL, Hopman VTP, van Putten JPM (1988) Gonococcal stress proteins, carriers of 'common bacterial antigens'? Abstract of 20th Lunteren Lectures on Molecular Genetics. Lunteren, Netherlands, pp 27–39

Peterson MG, Crewther PE, Thompson JK, Corcoran LM, Coppel RL, Brown GV, Anders RF, Kemp DJ (1988) A second antigenic heat shock protein of *Plasmodium falciparum*. DNA 7: 71–78

Plikaytis BB, Carlone GM, Pau CP, Wilkinson HW (1987) Purified 60-kilodalton *Legionella* protein antigen with *Legionella* specific and nonspecific epitopes. J Clin Microbiol 25: 2080–2084

Rees A, Scoging A, Mehlert A, Young DB, Ivanyi J (1988) Specificity of proliferative response of human CD8 clones to mycobacterial antigens. Eur J Immunol 18: 1881–1887

Richman SJ, Vedvick TS, Reese RT (1989) Peptide mapping of conformational epitopes in a human malarial parasite heat shock protein. J Immunol 143: 285–292

Rothstein NM, Higashi G, Yates J, Rajan TV (1989) *Onchocerca volvulus* heat shock protein 70 is a major immunogen in amicrofilaremic individuals from a filariasis-endemic area. Mol Biochem Parasit 33: 229–236

Sampson JS, O'Conner SP, Holloway BP, Plikaytis BB, Carlone CM, Mayer LW (1990) Nucleotide sequence of *htpB*, the *Legionella pneumophila* gene encoding the 58 kDa genus-common antigen. Submitted to Infect Immun

Scallon BJ, Bogitsh BJ, Carter CE (1987) Cloning of a *Schistosoma japonicum* gene encoding a major immunogen recognized by hyperimmune rabbits. Mol Biochem Parasit 24: 237–245

Searle S, Campos AJR, Coulson RMR, Spithill TW, Smith DF (1989) A family of heat shock protein 70-related genes are expressed in the promastigotes of *Leishmania major*. Nucleic Acids Res 17: 5081–5095

Selkirk ME, Denham DA, Partono F, Maizels RM (1989) Heat shock cognate 70 is a prominant immunogen in Brugian filariasis. J Immunol 143: 299–308

Sheppard M, Kemp DJ, Anders RF, Lew AM (1989) High level sequence homology between a *Plasmodium chabaudi* heat shock protein gene and its *Plasmodium falciparum* equivalent. Mol Biochem Parasit 33: 101–104

Shinnick TM (1987) The 65-kilodalton antigen of *Mycobacterium tuberculosis*. J Bacteriol 169: 1080–1088

Shinnick TM, Krat C, Schadow S (1987a) Isolation and restriction maps of the genes encoding five *Mycobacterium tuberculosis* proteins. Infect Immun 55: 1718–1721

Shinnick TM, Sweetser D, Thole J, van Embden JDA, Young RA (1987b) The etiologic agents of leprosy and tuberculosis share an immunoreactive protein antigen with the vaccine strain *Mycobacterium bovis* BCG. Infect Immun 55: 1932–1935

Shinnick TK, Vodkin MH, Williams JC (1988) The *Mycobacterium tuberculosis* 65-kilodalton antigen is a heat-shock protein which corresponds to common antigen and to the *Escherichia coli* GroEL protein. Infect Immun 56: 446–451

Shinnick TM, Plikaytis BB, Hyche AD, Van Landingham RL, Walker LL (1989) The *M. tuberculosis* BCG—a protein has homology with the *Escherichia coli* GroES protien. Nucleic Acids Res 17: 1254

Smejkal RM, Wolff R, Olenick JG (1988) *Leishmania braziliensis panamensis*: increased infectivity resulting from heat shock. Exp Parasitol 65: 1–9

Sorger PK, Pelham HRB (1987) Cloning and expression of a gene encoding hsc73, the major hsp70-like protein in unstressed rat cells. EMBO J 6: 993–998

Stover CK, Marana DP, Dasch GA, Oaks EV (1990) Molecular cloning and sequence analysis of an operon encoding the Sta58 major antigen of *Rickettsia tsutsugamushi*: Sequence homology and antigenic comparison to the 60 kDa family of stress proteins. Infect Immun 58: in press

Susek RE, Lindquist SL (1989) Hsp26 of *Saccharomyces cerevisiae* is related to the superfamily of small heat shock proteins but is without demonstrable function. Mol Cell Biol 9: 5265–5271

Taylor DW, Cordingley JS, Butterworth AE (1984) Immunoprecipitation of surface antigen precursors from *Schistosoma mansoni* messenger RNA in vitro translation products. Mol Biochem Parasitol 10: 305–318

Thole JER, Kevlen WJ, Kolk AHJ, Groothuis DG, Berwald LG, Tiesjema RJ, van Embden JDA (1987) Characterization, sequence determination, and immunogenicity of a 64-kilodalton protein of *Mycobacterium bovis* BCG expressed in *Escherichia coli* K-12. Infect Immun 50: 800–806

Thole JER, Hindersson P, de Bruyn J, Cremers F, van der Zee J, de Cock H, Thomassen H, van Eden W, van Embden JDA (1988) Antigenic relatedness of a strongly immunogenic 65 kDa mycobacterial antigen with a similarly sized ubiquitous bacterial common antigen. Micerobiol Pathogenesis 4: 71–83

Tsoufla G, Rook GAW, van Embden JDA, Young DB, Mehlert A, Isenberg DA, Hay FC, Lydyard PM (1989) Raised serum IgG and IgA antibodies to mycobacterial antigens in rheumatoid arthritis. Ann Rheum Dis 48: 118–123

VanBuskirk A, Crump BL, Margoliash E, Pierce SK (1989) A peptide binding protein having a role in antigen presentation is a member of the hsp70 heat shock family. J Exp Med 170: 1799–1809

Vodkin MH, Williams JC (1988) A heat shock operon in *Coxiella burnetti* produces a major antigen homologous to a protein in both mycobacteria and *Escherichia coli*. J Bacteriol 170: 1227–1234

Watson JD (1989) Leprosy: understanding protective immunity. Immunol Today 10: 218–221

Welch WJ, Mizzen LA, Arrigo AP (1989) Structure and function of mammalian stress proteins. In: Pardue ML, Feramisco JR, Lindquist S (eds) Stress-induced proteins. UCLA-ICN symposia on molecular and cellular biology, vol 76. Alan R Liss, New York, pp 187–202

Winfield JB (1989) Stress proteins, arthritis, and autoimmunity. Arthritis Rheum 32: 1497–1504

Young DB (1988) Stress-induced proteins and the immune response to leprosy. Microbiol Sci 5: 143–146

Young DB, Ivanyi J, Cox JH, Lamb JR (1987) The 65 kDa antigen of mycobacteria—a common bacterial protein? Immunol Today 8: 215–219

Young DB, Lathigra R, Hendrix R, Sweetser D, Young RA (1988a) Stress proteins are immune targets in leprosy and tuberculosis. Proc Natl Acad Sci USA 85: 4267–4270

Young DB, Mehlert A, Bal V, Mendez-Samperio P, Ivanyi J, Lamb JR (1988b) Stress proteins and the immune response to mycobacteria—Antigens as virulence factors? Antonie van Leeuwenhoek 54: 431–439

Young DB, Lathigra R, Mehlert A (1989) Stress-induced proteins as antigens in infectious diseases. In: Pardue ML, Feramisco JR, Lindquist S (eds) Stress-induced proteins. UCLA-ICN symposia on molecular and cellular biology, vol 76. Alan R Liss, New York, pp 275–285

Young RA (1990) Stress proteins and immunology. Ann Rev Immunol 8: 401–420

Young RA, Elliott TJ (1989) Stress proteins, infection, and immune surveillance. Cell 59: 5–8

Young RA, Bloom BR, Grossinsky CM, Ivanyi J, Thomas D, Davis RW (1985a) Dissection of *Mycobacterium tuberculosis* antigens using recombinant DNA. Proc Natl Acad Sci USA 83: 2583–2587

Young RA, Mehra V, Sweetser D, Buchanan T, Clark-Curtiss J, Davis R, Bloom BR (1985b) Genes for the major protein antigens of the leprosy parasite *Mycobacterium leprae*. Nature 316: 450–452

Stress Proteins, Autoimmunity, and Autoimmune Disease

J. B. WINFIELD and W. N. JARJOUR

1 Introduction

The venerable concept that mycobacteria and other infectious microorganisms cause chronic arthritis and persistent autoimmunity in genetically susceptible individuals is well supported, but the exact mechanisms by which this occurs are

Thurston Arthritis Research Center, Division of Rheumatology and Immunology, University of North Carolina, 932 FLOB 231HCB 7280, Chapel Hill, North Carolina 27599, USA

Current Topics in Microbiology and Immunology, Vol. 167
© Springer-Verlag Berlin · Heidelberg 1991

not well-understood (reviewed in) (SAAG and BENNETT 1987; OLDSTONE 1987, 1989; SHOENFELD and ISENBERG 1988; KEAT et al. 1989; ROSE 1989; VAUGHN 1990a, b). Recognition that the immunodominant 65-kD (hsp60)[1] and ~ 70-kD proteins of mycobacteria are homologues of *E. coli* GroEL and DnaK stress proteins is providing entirely new insight in this regard (SHINNICK et al. 1988; D. YOUNG et al. 1988; D.B. YOUNG et al. 1987), and has stimulated an extraordinary amount of productive investigation over the past few years (reviewed in D.B. YOUNG et al. 1988a; COHEN 1988, 1989b; ROOK 1988; KAUFMANN et al. 1989; VAN DEN BROEK 1989; VAN EDEN et al. 1989a, b; VAN EDEN and DE VRIES 1989; WINROW et al. 1990; DUBOIS 1989; R. YOUNG and ELLIOTT 1989; WINFIELD 1989; HURST 1990; VAN DER ZEE et al. 1990). Collectively, this work promises to answer many fundamental questions concerning the etiology of certain autoimmune disease.

Stress proteins are major immune targets in a broad spectrum of infectious diseases and have extremely close homologues in man. Because infection entails "stress' for both the microorganism and the host, e.g., during phagocytosis (CLERGET and POLLA 1990), increased synthesis and altered expression of extremely similar sets of autologous and foreign molecules occur at a time of active immune response, thereby placing stress proteins uniquely at the interface of tolerance and autoimmunity. The fact that infection with one microbe confers some degree of heightened immunity to many microbes because of immunological cross-reactions with microbial stress proteins suggests that too much emphasis has been placed in the past on establishing etiological relationships of specific microorganisms with individual autoimmune diseases. Clearly, the field already has advanced to the point where we now can focus on specific epitopes of microbe and host stress proteins in the genesis of autoimmune disease.

2 Initial Glues for Involvement of Stress Proteins in Autoimmune Disease

Initial support for a role of stress proteins in autoimmunity was obtained in four areas: (a) observations in the adjuvant (*Mycobacterium tuberculosis* in oil) arthritis model which suggested that autoreactive T cells triggered by mycobacterial antigens mediate persistent arthritis or protection from arthritis (reviewed in COHEN et al. 1985; VAN EDEN et al. 1987; ROOK 1988; WINFIELD 1989); (b) demonstrations that synovial fluid T cells from patients with rheumatoid arthritis and other types of arthritis proliferate in response to mycobacterial extracts or mycobacterial hsp60 (RES et al. 1988; GASTON et al. 1988); (c) data from three laboratories which suggested that T cells with $\gamma\delta$ receptors (BRENNER

---

[1] This protein is widely known as "65 K" or "mycobacterial 65 K." In this review, 65 K and all other *E. coli* groEL-equivalent proteins are referred to as hsp60

et al. 1988; STROMINGER 1989) exhibit special reactivity with mycobacterial antigens, including hsp60 (JANIS et al. 1989; O'BRIEN et al. 1989; HOLOSHITZ et al. 1989); and (d) reciprocal discoveries that patients with systemic autoimmune disease develop autoantibodies to stress proteins (MINOTA et al. 1988a, b; TAN et al. 1988; MULLER et al. 1988; BAHR et al. 1988; TSOULFA et al. 1988; WEISS et al. 1990) and that microbial stress proteins induce the formation of autoantibodies (MATTEI et al. 1989). Two important studies provided insight into the mechanisms by which a mycobacteria-induced immune response could result in autoreactive tissue injury. The recombinant hsp60 stress protein of *M. bovis* BCG/*M. tuberculosis* was shown to stimulate CD4$^+$ cytotoxic lymphocytes to lyse not only purified protein derivative (PPD) or hsp60-pulsed monocytes, but nonpulsed or irrelevant antigen-pulsed monocytes as well (OTTENHOFF et al. 1988). This observation suggested that mycobacterial hsp60 cross-reactive autologous antigen is expressed on uninfected cells, thereby rendering them susceptible to immunological attack. The scope of such T cell autoreactivity was further extended by the observation that major histocompatibility complex (MHC) class I- and class II-negative Schwann cells be stimulated with IFN-γ to express MHC class I molecules which were capable of presenting immunodominant *M. leprae* antigens, resulting in specific lysis of Schwann cells by CD8$^+$ T cells (STEINHOFF and KAUFMANN 1988). This type of tissue injury previously had been thought to be due to nonspecific bystander effects.

Extraordinary progress has been made in the accumulation of stress protein structural data. A constitutively expressed mitochondrial protein was identified recently as the human and rodent homologue of GroEL and mycobacterial hsp60 (MIZZEN et al. 1989; DUDANI and GUPTA 1989; MCMULLIN and HALLBERG 1988; JINDAL et al. 1989). Human hsp60 exhibits >50% sequence homology and immunological cross-reactivity with GroEL, mycobacterial hsp60, and the ribulose-1,5-bisphosphate carboxylase/oxygenase (Rubisco) subunit binding protein of plant chloroplasts (LANDRY and BARTLETT 1989; JINDAL et al. 1989; DUDANI and GUPTA 1989). cDNA clones and sequences are now available for hsp60 and for many other stress proteins, including: DnaJ, DnaK, and GroEL (*E. coli*) and related proteins in mycobacteria, yeast, and mammals, small stress proteins, and several members of the hsp70 and hsp90 families (SADLER et al. 1989; LATHIGRA et al. 1988; BARDWELL and CRAIG 1984; R. YOUNG et al. 1985; CHANG et al. 1987; MUSTAFA et al. 1988; REQUENA et al. 1989; READING et al. 1989; FARRELLY and FINKELSTEIN 1984; VAN EDEN et al. 1988; YAMAZAKI et al. 1989; REBBE et al. 1989; JINDAL et al. 1989). Such information is facilitating a rapidly expanding body of data concerning the epitopes with which B and T cells react (e.g., VAN EDEN et al. 1988; MCMULLIN and HALLBERG 1988; ANDERSON et al. 1988; OFTUNG et al. 1988; MEHRA et al. 1986; MUNK et al. 1988, 1989; ROBAYE et al. 1989; RICHMAN et al. 1989; LAMB et al. 1989; THOLE et al. 1988; MELANCON KAPLAN et al. 1988; ANDERSON et al. 1988; D.B. YOUNG et al. 1988b; BRETT et al. 1989). For example, peptides synthesized according to conserved mycobacterial and mammalian hsp60 180–196 sequences strongly stimulate some neonatal mouse thymocyte hybridomas (Born et al. 1990b). The core epitope recognized by arthritogenic

$\alpha\beta$ T cells has been mapped to the non-conserved 180–188 region of mycobacterial hsp60 (VAN EDEN et al. 1988).

3 Stress Proteins and the Immune System

Although their function was quite mysterious as recently as 5 years ago, it is now clear that stress proteins are fundamentally important in the biology of the cell (reviewed in SCHLESINGER 1986; LINDQUIST 1986; BOND and SCHLESINGER 1987; WELCH et al. 1989; ELLIS and HEMMINGSEN 1989), in the immune system general (KAUFMANN 1990), and in inflammation (POLLA 1988; KÖLLER et al. 1989; KUBO et al. 1985). grp78 (BiP) and hsp70 stress proteins function as chaperones in the translocation of peptides across membranes and in their assembly into larger molecules and molecular complexes (ELLIS and HEMMINGSEN 1989), e.g., in the formation of immunoglobulin molecules (HAAS and WABL 1983; MUNRO and PELHAM 1986; HAAS and MEO 1988) and in antigen processing and presentation to T cell receptors (TCRs) (LAKEY et al. 1987; MILLER et al. 1989; VANBUSKIRK et al. 1989; TOMASOVIC et al. 1989; PARHAM 1989; TOWNSEND et al. 1989), respectively. GroEL-related stress proteins (hsp60 homologues) are chaperones as well (LECKER et al. 1989; LANDRY and BARTLETT 1989; HEMMINGSEN et al. 1988), raising the possibility that hsp60 could also play a role in antigen processing. Of interest in this regard is the recent identification of a mammalian cytoplasmic protein, TCP-1, encoded by a gene within the mouse t-complex (encompassing the MHC) as a hsp60 homolog (Gupta 1990). Lymphocyte homing (SIEGELMAN et al. 1986), resistance to target cell lysis (WILSON et al. 1989; RENKONEN et al. 1988), tumor antigenicity (ULLRICH et al. 1986; STRAHLER et al. 1990), lymphocyte and macrophage activation (HAIRE et al. 1988; POLLA 1988; GRANELLI PIPERNO et al. 1986; FERRIS et al. 1988; MAHER and PASQUALE 1989), $\gamma\delta$ T cell selection (see below), and immune surveillance (BORN et al. 1990) are other immunological functions that may be mediated or modulated by stress proteins. In addition, genes for hsp70 have been located in the class III region of the human and rat MHC (SARGENT et al. 1989; WURST et al. 1989), and the gene for hsp84 is linked to the H-2 complex in the mouse (ROMANO et al. 1989).

4 Concepts of Immune System Physiology and Autoimmunity

To provide a framework for considering the role of stress proteins in autoimmunity, current thought concerning the physiology of immune system function and the genesis of autoimmune disease in genetically predisposed individuals is reviewed briefly. Relevant concepts (which reflect the authors' bias) include: (a) COUTINHO's argument that autoreactivity of B and T lymphocytes is physiological and, not-withstanding the importance of thymic and peripheral deletion and veto effects, central to self–nonself discrimination (COUTINHO 1989; COUTINHO and BANDEIRA 1989; PEREIRA et al. 1989); (b) the idea that humoral and cellular immune responsiveness is, in large part, evolutionarily "set" at birth to protect the host from infection and that this is responsible for the nonrandom nature of autoimmunity (COHEN 1989a; VAN ROOIJEN 1989; SHOENFELD et al. 1989); and (c) Born's proposal that conserved stress protein epitopes constitute an important antigen element which shapes the immune system (BORN et al. 1990).

4.1 Selection of T Cells

Negative selection in the thymus of $\alpha\beta$ and $\gamma\delta$ T cells that recognize self-MHC molecules in combination with other self-molecules is a major mechanism underlying self-tolerance (reviewed in Schwartz 1989). Because of limitations in peptide presentation, negative selection in the thymus may not be complete, however. This allows peripheral T cell recognition of some autologous peptides that are not presented in sufficient density by MHC-expressing cells in the thymus (non-physiologic presentation) (SCHILD et al. 1990). Positive selection of $\alpha\beta$ T cells by intrathymic expression of appropriate MHC restriction elements skews the T cell repertoire toward recognition of foreign antigens in the context of self-MHC molecules. Less polymorphic $\gamma\delta$ T cells are positively selected, at least in the mouse, by restricted sets of as yet undefined self-peptides (LAFAILLE et al. 1990). Some of these intrathymic peptides may be related to microbial antigens (? stress proteins), as suggested by the high proportion of neonatal thymocyte hybridomas that recognize mycobacterial antigens, including hsp60 and a conserved hsp60 peptide (O'BRIEN et al. 1989; BORN 1990b). Neonatal thymocyte hybridomas also exhibit unusual, TCR-dependent "self-stimulatory" activity, probably due to recognition of myco-bacterial hsp60 cross-reactive autologous stress protein peptides (O'BRIEN et al. 1989). These and other data have been synthesized into an interesting schema (BORN et al. 1990) wherein at least some $\gamma\delta$ cells recognize conserved epitopes of autologous stress protein peptides presented by simple, less polymorphic antigen-presenting molecules (CD1, TL, Qa, reviewed in STROMINGER 1989; PORCELLI et al. 1989). BORN and colleagues propose that $\gamma\delta$ cells which recognize stress proteins represent a "rapid-response" first line of defense against infection. With stress or other stimuli, autologous stress proteins are "presented" on cell surfaces to special subsets of $\gamma\delta$ cells, which respond by mediating cellular activation, immune surveillance, and regulation of lymphocyte growth and differentiation. Cross-reactions of autologous stress proteins with bacteria may lead to a break in tolerance, expansion of autoreactive T cells that recognize stress proteins, and autoimmune-mediated cell and tissue injury. Consistent with BORN's model are data showing that brief heat shock of murine T cells induces selective proliferation of $\gamma\delta$ cells (RAJASEKAR et al. 1990). In these experiments, enrichment of $\gamma\delta$ cells following heat shock is increased by preexposure of the cells to mycobacterial antigens or by culturing lymphocytes with heat-shocked syngeneic bystander cells. Taken together, the accumulated data suggest a possible dual role of $\gamma\delta$ cells [and, by analogy, double-negative CD4$^-$/CD$^-$ $\alpha\beta$ cells (PORCELLI et al. 1989)] in host defense against infectious microorganisms and, in genetically susceptible individuals, in the induction and/or amplification of T cell autoreactivity in autoimmune disease.

Stress proteins obviously do not explain the entire interface between bacteria and host in thymocyte ontogeny, tolerance, and autoimmunity, however. For example, self-protein "superantigens" related to staphylococcal enterotoxins, such as Mls gene products (JANEWAY et al. 1988, 1989), appear to play a role in positive and negative selection of T cells bearing certain V_β receptors and, when administered at birth to mice, can induce tolerance (FLEISCHER 1989; JANEWAY et al. 1989; WHITE et al. 1989). A compelling argument has been made that bacterial superantigens are fundamentally important in generating T cell help during the development of B cell autoantibody responses in systemic autoimmune diseases such as SLE (Friedman et al. 1990). Furthermore, PPD and mycobacterial hsp60 are not the major activators, at least in PPD-nonimmune individuals, of mycobacteria-reactive $\gamma\delta$ T cells in the peripheral blood of normal individuals (KABELITZ et al. 1990). Such observations indicate that mycobacterial polysaccharide or lipids also are important in stimulating $\gamma\delta$ cells in man. Nevertheless, it is clear that selection of $\alpha\beta$ and $\gamma\delta$ T cells is strongly influenced by self-antigens which are related to those of infectious pathogens, including stress proteins, as part of an evolutionarily conserved mechanism to protect the host from infection.

4.2 Natural Antibodies

The development of natural antibodies and the selection of $\gamma\delta$ cells exhibit interesting parallels. Like $\gamma\delta$ cells, neonatal B cell hybridomas secrete germ-lined encoded polyspecific antibodies that frequently recognize self-antigens (TERNYNCK and AVRAMEAS 1986; TRON and BACH 1989). As formulated by COUTINHO, these IgM antibodies exhibit high idiotypic connectivity which contributes importantly to the development of the adult B cell repertoire (COUTINHO 1989; COUTINHO and BANDEIRA 1989; PEREIRA et al. 1989). Whether the repertoire of IgM natural antibodies, as with $\gamma\delta$ cells, is biased toward reactivity with stress proteins is unknown, but it is reasonable that such should be the case. A second parallel between neonatal B cells and neonatal $\gamma\delta$ cells is the capacity of both to "self-stimulate" (COUTINHO 1989; O'BRIEN et al. 1989). IgM antibodies stimulate an extremely rapid expansion of B cells in the first few weeks after birth through interactions with as yet undefined non-Ig molecules and surface Ig (idiotype/anti-idiotype interactions). As has been proposed for $\gamma\delta$ cells (BORN et al. 1990), this would enable both a "preselection" and a "recruitment" of B cells expressing repertoires optimum for host defense. Again, it is reasonable to suggest that the dominant specificities of such B cells involve stress proteins and that similar mechanisms obtain in the generation of B cells in adults. Initial development of the B cell repertoire probably is T cell independent, but at some point T cells become involved in an equilibrated "global immune network" involving T cells, B cells, and self-antigens (? including conserved epitopes of autologous stress proteins) (COUTINHO 1989).

4.3 Immune Network

The autoreactive network functions to sustain and select an appropriate pool of "connected" B and T cells as the immune system evolves throughout life. If newly formed resting lymphocytes do not encounter a self specificity, they die. Stress proteins cross-reactive with mycobacterial hsp60 may represent one of the important autologous antigenic stimuli in this regard. For example, the V- and J-region gene segments expressed by murine $\gamma\delta$ cells become more heterogeneous when $\gamma\delta$ cells are stimulated to proliferate by heat shock (RAJASEKAR et al. 1990). COUTINHO proposes that the recognition structures of some of these network-connected lymphocyte clones, while exhibiting low-level reactivity with self, are better suited to respond to novel nonself shapes. Such novel shapes certainly must include nonconserved or species-specific stress protein epitopes in bacteria, as well as autologous peptides which had not been presented "physiologically" to T cells in the thymus. Encounters with foreign antigen, or the development of pathologic conditions which lead to the expression of "non-physiologic" combinations of MHC molecules and self-peptides, result in rapid expansion of "disconnected" B and T cells, production of high levels of antibodies with greater affinity for the novel antigen, and differentiation of T cells to effector functions in what classically has been considered to be an immune response. Once the novel antigen has been cleared, such disconnected clones virtually disappear because only self, not foreign, molecules remain to select lymphocytes.

Regarding the implications of the above for autoimmunity, COUTINHO views organ-specific autoimmunity as representing a localized defect in connectivity such that autoreactive clones, being disconnected, are able to proliferate and mutate. Generalized (systemic) autoimmunity would represent a more global defect in V-region connectivity and/or regularoty mechanisms. Consistent with this thesis are recent data showing that the number of preactivated and MHC class II-restricted autoreactive T cells in peripheral blood of patients with rheumatoid arthritis are dramatically increased relative to the levels in healthy individuals (SCHLESIER et al. 1989). Similarly, peripheral blood T cells from patients with inflammatory arthritis proliferate strongly, in an antigen-specific fashion, to autologous synovial fluid non-T cells in the absence of exogenous antigen or mitogen (LIFE et al. 1990). $\gamma\delta$ cells appear to represent a sizable fraction of such cells in arthritic joints (DE MARIA et al. 1987; HOLOSHITZ et al. 1989;

BRENNAN et al. 1988; PORCELLI et al. 1989), at other sites of ongoing autoimmune tissue injury, and in human infectious disease lesions (MODLIN et al. 1989; HAREGEWOIN et al. 1989). Increased numbers of $CD4^-/CD8^-$ $\alpha\beta$ T cells in the peripheral blood of patients with systemic lupus erythematosus (SLE) may exemplify an analogous phenomenon (SHIVAKUMAR et al. 1989).

4.4 Generation of Pathological T Cell and Antibody Autoreactivity

Exactly why abnormal levels of high affinity IgG antibodies to self-constituents and aggressive T cell autoreactivity develop in certain individuals is unknown, but clearly these processes utilize germ line genes and are antigen driven. Some important elements include:

1. The presence of "permissive" disease-related MHC class I and class II genes (TODD et al. 1988; ARNETT et al. 1989; NEPOM 1989a, b; GREGERSEN 1989; HOLMDAHL et al. 1989; Nepom et al. 1989; CACCIA and MAK 1988)
2. MHC class III genes and complement polymorphisms and deficiencies (ATKINSON 1988; RITTNER and SCHNEIDER 1989)
3. Aberrant or increased expression of HLA molecules able to present antigen at sites of autoimmune tissue injury (FELDMANN 1989; BOTTAZZO et al. 1983; MATIS et al. 1983)
4. Non-MHC genes that predispose to generalized autoimmunity (SLOR et al. 1989; WINCHESTER and NONEZ-ROLDAN 1982)
5. Inherited differences in immunoglobulin genes (e.g., CHEN et al. 1989)
6. TCR genes (HAQQI et al. 1989; BANERJEE et al. 1988; KOFLER et al. 1989; THEOFILOPOULOS and KOFLER 1989)
7. Cytokines (LIPSKY 1989; BALKWILL and BURKE 1989; WILDER et al. 1989; LAFYATIS et al. 1989; KUMKUMIAN et al. 1989; RITCHLIN and WINCHESTER 1989; ZIFF 1989; LIPSKY et al. 1989; KROEMER and WICK 1989)
8. Natural killer cells (GRUNEBAUM et al. 1989)
9. Ly-1$^+\beta$ cells (HERZENBERG et al. 1989; VAN ROOIJEN 1989)
10. Hormonal factors (TALAL 1989)
11. Immunoregulatory abnormalities (MORIMOTO et al. 1988; SINCLAIR and PANOSKALTSIS 1988)
12. Triggering or inciting agents (ROSE 1989; VAUGHN 1990).

A long list of potential triggers has been assembled, e.g., foreign antigens of viruses or bacteria, release of sequestered antigen, altered self-antigen, self-antigen + increased MHC molecules to present them, anti-idiotypes (ZANETTI et al. 1984; SHOENFELD and ISENBERG 1989; SHOENFED et al. 1989), polyclonal B cell activators (DZIARSKI 1988), bacterial superantigens, and chronic Inflammation itself (SARVETNICK et al. 1988). Molecular mimicry (Froude et al. 1989; Fujinami and Oldstone 1989) (which theoretically could involve any of the "shapes" recognized by MHC molecules and by specific and non-specific receptors on B and T cells), the immunogenic particle concept advanced by Tan to explain characteristic "linked sets" of autoantibodies in certain diseases (Tan et al. 1988), bacterial superantigens which induce intense cognate T-help/B-cell interactions (Friedman et al. 1990), and the non-physiologic presentation of autologous peptides to self-reactive T cells which escaped negative selection in the thymus (Schild et al. 1990) have been invoked as principal reasons for the failure of the immune system to discriminate self from non-self in overt autoimmune disease. With respect to stress proteins, most attention has been focused on molecular mimicry between prokaryotes and eukaryotes, but it should not be forgotten that immunological cross-reactions exist between stress proteins and virus proteins (SHESHBERADARAN and NORRBY 1984).

5 Stress Proteins in Animal Models of Human Autoimmune Disease

5.1 Inflammatory Arthritis

Several lines of evidence from the adjuvant arthritis model in rats support an etiological relationship between mycobacterial antigens, or mycobacteria-related antigens of other infectious microorganisms, and chronic inflammatory arthritis in man. As reviewed previously (COHEN et al. 1985; VAN EDEN et al. 1987; ROOK 1988; COHEN 1988; VAN EDEN et al. 1989a, b), a T cell line (A2), derived from *M. tuberculosis* antigen-containing cultures of lymph node cells of rats primed with Freund's complete adjuvant, transfers arthritis when injected intravenously into irradiated, syngeneic rats that have never been exposed to *M. tuberculosis*. Injection of A2 cells into nonirradiated rats induces resistance to adjuvant arthritis. A clone from the A2 cell line (A2b) possesses even greater arthritogenic activity and exhibits immunological reactivity in vitro and in vivo to *M. tuberculosis* and cartilage proteoglycan. Administration of a second A2 clone (A2c) does not cause arthritis, but instead confers protection against adjuvant arthritis and accelerated remission in rats with already established disease. Using clones A2b and A2c to screen mycobacterial antigens expressed in *E. coli*, hsp60 was identified as a critical antigen in the adjuvant arthritis model (VAN EDEN et al. 1988).

Immunization of rats with mycobacterial hsp60 induces resistance to adjuvant arthritis following subsequent immunization with Freund's complete adjuvant (VAN EDEN et al. 1988; BILLINGHAM et al. 1990). Surprisingly, pre-administration of mycobacterial hsp60 also protects rodents from developing arthritis induced by completely unrelated agents, i.e., pristane-induced arthritis (THOMPSON et al. 1990), streptococcal cell wall-induced arthritis (VAN DEN BROEK et al. 1989; VAN DEN BROEK 1989), and to a lesser extent, CP20961 (Avridine)—and type II collagen-induced arthritis (BILLINGHAM et al. 1990). Mycobacterial hsp60-injected rats also exhibit a suppressed proliferative response to concanavalin A, but arthritis induced by zymosan and delayed-type hypersensitivity reactions to unrelated protein are not affected (VAN DEN BROEK et al. 1989). The mechanisms underlying these phenomena are unknown. But given the interaction of $\gamma\delta$ T cells with stress proteins discussed above, it is possible that administration of mycobacterial hsp60 is inducing some type of suppressive immunoregulation which prevents the clonal proliferation of autoreactive T cells.

A cross-reaction between an acetone-precipitated fraction of *M. tuberculosis* and cartilage proteoglycan has been reported in adjuvant arthritis (VAN EDEN et al. 1985, 1989b) and in rheumatoid arthritis (HOLOSHITZ et al. 1986). T cell clones obtained from rats with adjuvant arthritis were shown subsequently to recognize a mycobacterial hsp60 nonapeptide (amino acid position 180–188) having minor sequence similarity with the link protein of rat proteoglycan (VAN EDEN et al. 1988). The relevance of these findings to human arthritis and to specific autoimmune attack against cartilage is subject to debate, however. First, the 180-188 region of mycobacterial hsp60 shows little similarity with human hsp60. Second, other antigens share this sequence similarity with hsp60, e.g., the antigen BLPF-1 of Epstein-Barr virus and HLA-DQ3. Third, other mycobacterial hsp60 epitopes, or even mycobacterial proteins other than hsp60, may be as important in generating T cell autoreactivity in adjuvant arthritis (D.B. YOUNG et al. 1988a; GASTON et al. 1990; DEJOY et al. 1989) and in rheumatoid arthritis (GASTON et al. 1990). Nonetheless, further studies of the vaccination and therapeutic potential of the 180–188 nonapeptide (Thr-Phe-Gly-Leu-Gln-Leu-Glu-Leu-Thr) indicate that, when preadministered intraperitoneally, it reduces the incidence and severity of adjuvant arthritis in a fashion analogous to that mediated by mycobacterial hsp60 in adjuvant arthritis, streptococcal cell well-induced arthritis, and pristane-induced arthritis (YANG et al. 1990). This activity was abrogated completely by deletion of the N-terminal threonine. Moreover, a related peptide that contains conserved sequences of mycobacterial hsp60 (position 180–196) stimulates murine neonatal $\gamma\delta$ thymocyte hybridomas (BORN 1990b; see O'BRIEN et al. 1989). Collectively, these diverse observations suggest that mycobacterial hsp60, while just one of many inducers of arthritis, may react with $\gamma\delta$ cells in a relatively unique way and may induce important immunoregulatory effects in different types of arthritis. An additional clue to the possible involvement of autologous hsp60 peptides in this regard is the intriguing observation that administration of gold to donor rats inhibits the passive transfer of adjuvant arthritis (CANNON and McCALL 1989). Gold is a stress protein inducer (CALTABIANO et al. 1986, 1988) and is one of the oldest remedies for rheumatoid arthritis.

5.2 Diabetes

A beta cell antigen cross-reactive with mycobacterial hsp60 has been implicated in the induction of autoimmune diabetes in the nonobese diabetic mouse (ELIAS et al. 1990). In this model, onset of beta cell destruction closely parallels the development of T cells reactive with hsp60. Release into the circulation of a beta cell protein cross-reactive with hsp60 is followed by the appearance of anti-hsp60 autoantibodies, anti-insulin antibodies, and anti-idiotype antibodies. Anti-hsp60 T cell clones are able to transfer the disease to young mice, and administration of hsp60 antigen either induces diabetes (hsp60 + incomplete Freund's adjuvant) or vaccinates (hsp60 + saline) against diabetes. Preliminary data suggest that hsp60 may play a similar role in spontaneously diabetic BB rats and in early human diabetes (ELIAS et al. 1990), although virus proteins may also be important in the latter situation (KAROUNOS et al. 1990). Of unknown relevance to the role of stress proteins in diabetes is a recent report that insulin induces the expression of the human *hsp 70* gene (TING et al. 1989).

5.3 Trachoma

The chronic eye inflammation of blinding trachoma which accompanies infection with *Chlamydia trachomatis* has the characteristics of a delayed-type hypersensitivity response. A 57-kD homologue of *E. coli* GroEL in chlamydia is strongly implicated in the eye inflammation of this disease (MORRISON et al. 1989). The pathogenesis of trachoma probably can be generalized to chlamydial disease in the joints (KEAT et al. 1987, 1989; HOUGH and RANK 1989) and the urogenital system and to tissue injury accompanying other infections, e.g., syphilis, tuberculosis, and leprosy.

6 Stress Proteins and Inflammatory Arthritis in Man

The provocative observations implicating mycobacterial hsp60 in adjuvant arthritis in rats impelled intense efforts in many laboratories to find similar relationships in rheumatoid arthritis and other types of inflammatory arthritis in man. Since chronic arthritis persists in the absence of putative microbial triggers, one line of investigation has focused on analyses of stress proteins in joint tissues and cells. A caveat to be recalled in interpreting such studies, however, is that microbial stress proteins may indeed still be present in some types of chronic arthritis. For example, application of polymerase chain reaction technology recently demonstrated *Borrelia burgdorferi* in the joints of patients with chronic, antibiotic-resistant Lyme disease (NIELSEN and PETER 1990).

6.1 Localization of Stress Proteins in Cells
and Tissues of the Joint

In studies of tissue specimens, a mouse monoclonal antibody (ML30) raised against mycobacterial hsp60 detects high levels of the human homologue of hsp60 in the cartilage/pannus junction and in rheumatoid nodules of patients with rheumatoid arthritis, but not in normal tissues or in several other types of chronic tissue inflammation (KARLSSON-PARRA et al. 1990). Using different antibody probes, other investigators have found no hsp60, increased hsp60, and increased intracellular hsp70 and hsp90 in synovial cells and in synovial fluid cells from patients with rheumatoid arthritis, other types of inflammatory arthritis, and even osteoarthritis (MCLEAN et al. 1988; MAPP et al. 1989; CHERRIE et al. 1989; KAUFMANN 1990). Chondrocytes from patients with osteoarthritis, not unexpectedly, constitutively express hsp70 and hsp80 (KUBO et al. 1985). Preliminary data suggest that hsp60 and members of the hsp70 and hsp90 families are expressed on synovial fluid cells obtained from patients with arthritis (JARJOUR et al. 1989, unpublished results). This latter issue obviously is most relevant to the nature of T cell autoreactivity in inflammatory joint disease.

6.2 T Cell Reactivity with Stress Proteins

Synovial fluid T cells exhibit an increased proliferative response to recombinant mycobacterial hsp60 and to an acetone-precipitable fraction of *M. tuberculosis* in patients with early rheumatoid arthritis (MCMULLIN and HALLBERG 1988). This observation has been confirmed and extended to patients with reactive arthritis and spondyloarthropathy, where increased synovial T cell reactivity to gram-negative bacteria is evident as well (GASTON et al. 1988, 1989). Proliferative responses to gram-negative bacteria and to mycobacterial hsp60 are highly correlated in rheumatoid arthritis, but not in reactive arthritis. Synovial fluid T cell reactivity is higher than that of peripheral T cells, suggesting that an intra-articular antigenic stimulus is present. However, when peripheral blood T cells from patients with inflammatory arthritis are cultured with autologous synovial fluid non-T cells in the absence of exogenous antigen or mitogen, a strong, antigen-specific proliferative response can be demonstrated (LIFE et al. 1990). Interestingly, synovial fluid T cells do not show this "spontaneous" reactivity, although they proliferate when cultured with synovial fluid non-T cells + bacteria or mycobacterial hsp60.

The mechanisms underlying these unusual phenomena are unknown, but are likely to be different in different types of arthritis. A common denominator for the enhanced capacity of synovial fluid antigen-presenting cells to stimulate T cells undoubtedly are the effects of cytokines to upregulate non-T cells with respect to expression of peptide-presenting molecules, adhesion molecules, and/or autologous stress proteins or stress protein peptides. The heightened response of synovial fluid T cells in patients with arthritis in these and earlier experiments (FORD et al. 1985; SIGNAL et al. 1986) may be related to intra-articular localization of bacterial antigens (GRANFORS et al. 1989) which attract antigen-specific T cells into the joint, where they undergo clonal expansion. This explanation is reasonable in patients with acute reactive arthritis, where microbial antigens can persist for considerable periods of time after the actual infection is cleared. In spondyloarthropathies, it has been proposed that bacterial superantigens stimulate T cell stimulate with shared Vβ receptors to such a degree that they become capable of sustaining an autoaggressive response in the absence of the initiating bacterial superantigen (WHITE et al. 1989). The accumulated data implicating mycobacterial hsp60 as a key inciting antigen in the adjuvant arthritis model and as a stimulator of neonatal $\gamma\delta$ thymocytes in the mouse (O'BRIEN et al. 1989) and of $\gamma\delta$ cells in the joints of patients with rheumatoid arthritis (HOLOSHITZ et al. 1989) would suggest that hsp60 can initiate and sustain T cell autoreactivity in human arthritis. Alternate mechanisms may also obtain, however. For example, the DnaJ stress protein of *E. coli* contains an amino acid sequence (Glu-Lys-Arg-Ala-Ala) found in the MHC DRβ third hypervariable region susceptibility locus for rheumatoid arthritis (D. CARSON, personal communication). This provocative observation raises entirely new questions regarding gram-negative bacteria and their stress proteins in the etiology of rheumatoid arthritis.

Two recent reports provide indirect support for the potential involvement of autologous hsp60 in T cell autoreactivity, at least with respect to effector function. In the mouse, CD8$^+$ $\alpha\beta$ cytolytic T cells (CTLs), raised against tryptic fragments of mycobacterial hsp60, lyse bone marrow-derived macrophages fed hsp60 peptides (as expected), but also lyse, in an MHC class I-restricted fashion, naive macrophages stimulated with IFN-γ or infected with cytomegalovirus (KOGA et al. 1989). These data suggest that stress and other stimuli induce the cell surface expression of autologous, mycobacterial hsp60 cross-reactive stress protein epitopes that render the targets susceptible to lysis by CTLs. Similar studies have been carried out in man (MUNK et al. 1989). Thus, peripheral blood CTLs activated in vitro with killed *M. tuberculosis* are cytotoxic for autologous monocytes primed with killed mycobacteria, with tryptic fragments of mycobacteria, with recombinant mycobacterial hsp60, with synthetic peptides corresponding to fully conserved sequences shared by mycobacterial hsp60 and its human homologue, and, in some cases, *unprimed monocytes*. The epitopes here are recognized in an MHC class II-restricted fashion. Taken together, these data clearly suggest that autologous stress protein epitopes can be expressed on the surface of target cells under certain circumstances and that healthy individuals have T cells in their circulation which recognize strictly autologous stress protein epitopes.

The observations of MUNK and colleagues (1989) implicate autologous stress protein epitopes in a confirmation of COUTINHO's concept of physiological T cell autoreactivity (COUTINHO 1989), but do not clarify the issue of which hsp60 epitopes are recognized in patients with rheumatoid arthritis. As discussed above with respect to animal models of arthritis, considerable attention has been focused on an epitope located in a nonapeptide at position 180–188 of mycobacterial hsp60. It was reported recently, however, that hsp60-reactive T cell clones from the synovial fluid of a patient with acute arthritis exclusively recognize a different peptide in the NH$_2$ terminus (GASTON et al. 1990). In analyses of synthetic peptides of the entire hsp60 molecule, no other epitopes could be shown to react with these clones. This dominant epitope does not contain conserved sequences, suggesting that T cells which contribute to autoimmunity do not necessarily have to be directed at a shared epitope.

6.3 Other Potential Contributions of Stress Proteins to Rheumatoid Disease

Given the extremely broad reach of stress proteins in basic cell biology and physiology, it is likely that stress proteins will be found to contribute to the clinical manifestations of arthritis and other forms of chronic arthritis in unexpected ways. For example, grp78 was recently identified as a binding protein for substance P (OBLAS et al. 1990), a neurotransmitter that has been implicated in joint inflammation (PAYAN 1989; LEVINE et al. 1987). Similarly, hsp70 is heavily localized in atherosclerotic plaques (BERBERIAN et al. 1990), and it is conceivable that hsp70 or other stress proteins play some role in the development of the extra-articular cardiovascular manifestations of rheumatoid arthritis (FERRANS and RODRIGUEZ 1985) and, perhaps, vasculitis generally.

7 Stress Proteins and Other Autoimmune Diseases

Except for some beginning work in diabetes (see above), the discovery of autoantibodies to stress proteins in SLE and related disorders (see below), and speculation that stress proteins may be important in myocarditis (Maisch 1989a–c), essentially no data regarding the role of stress proteins in other organ-specific and systemic autoimmune diseases are available. Given the current high level of interest in stress proteins vis-à-vis autoimmunity, however, this situation should be rectified rapidly. Prime candidates for investigation in this regard are autoimmune skin diseases, autoimmunity in the elderly, and autoimmune neuritis. For example, could polymorphic light eruption be a consequence of ultraviolet radiation-induced increases in the levels of hsp70 (BRUNET and GIACOMONI 1989) and heme oxygenase, a 32-kD stress protein (BOYLSTON 1987)? Could the decreased capacity for stress protein induction in old cells or old people (DEGUCHI et al. 1988; LIU et al. 1989a, b; FAASSEN et al. 1989) by related causally to decreased physiological "connectivity" and increased "disconnected" T cell autoreactivity in the elderly? Finally, the recent observation that complete Freund's adjuvant induces an experimental allergic neuritis reminiscent of adjuvant arthritis (MIZISIN et al. 1987) should excite an interest in stress proteins among neuroimmunologists.

8 Autoantibodies to Stress Proteins in Systemic Lupus Erythematosus and Other Disorders

The evidence linking mycobacterial infection with humoral autoimmunity has been reviewed previously (SHOENFELD and ISENBERG 1988). Thus, hypergammaglobulinemia is associated with active tuberculosis, and tubercle bacilli are extremely potent adjuvants for induction of polyclonal B cell activation. Patients with pulmonary tuberculosis develop rheumatoid factor and many of the same antinuclear autoantibodies characteristic of patients with SLE. Some murine and human monoclonal antibodies to DNA bind to mycobacteria; conversely, some monoclonal antibodies to mycobacteria bind to DNA and express the 16/6 anti-DNA idiotype. Persistent arthritis has been reported in patients given BCG immunotherapy. Collectively, such observations are consistent with the concept of molecular mimicry (OLDSTONE 1987) between mycobacterial antigens and host constituents in the genesis of some forms of autoimmunity, if not autoimmune disease. Although mycobacterial carbohydrate and lipid antigens clearly are important elements in this regard, certain evidence suggests a contribution by stress proteins as well.

The behavior of hsp70 and ubiquitin in the cell raises the possibility that stress proteins may be key ingredients in the "immunogenic particle" concept (TAN et al. 1988) of the origin of antinuclear and other autoantibodies, as has been suggested previously (MULLER et al. 1988; INOUE 1989). Thus, hsp70 associates with proteins in the nucleus and nucleolus in a cell cycle-dependent fashion (MILARSKI et al. 1989), migrates among different cell compartments, and binds to a variety of nuclear and cytoplasmic proteins in virus-infected cells (WELCH and SUHAN 1986; WELCH et al. 1985; SAWAI and BUTEL 1989; PALLAS et al. 1989; LA THANGUE and LATCHMAN 1988; NEVINS 1982; WHITE et al. 1988; JAY et al. 1978; OPPERMAN et al. 1981; SCHUH et al. 1985; COLLINS and HIGHTOWER 1982) and in cells exposed to other stressful stimuli. Similarly, ubiquitin forms particles ubiquitously as part of a mechanism for the clearance of denatured proteins. Perhaps also contributing to the induction of anti-stress protein autoantibodies is the elevation of stress proteins in peripheral blood mononuclear cells from patients with SLE, but not rheumatoid arthritis (DEGUCHI et al. 1987; LATCHMAN et al. 1990; NORTON et al. 1989). This phenomenon probably reflects endogenous B and T cell activation due to cytokine stimulation in vivo (YUFU et al. 1989; FERRIS et al. 1988; HAIRE et al. 1988). Conversely, a reduced capacity for induction of stress protein synthesis, as reported for aging cell lines (LIU et al. 1989a, b) and for lymphocytes from aged donors (DEGUCHI et al. 1988; FAASSEN et al. 1989), conceivably contributes to diminished network connectivity and increased autoantibody formation in the elderly, as discussed above.

8.1 Systemic Lupus Erythematosus

Histone H2B is a well-known autoantigen in SLE and related disorders (TAN et al. 1988), but only recently was defined as a stress protein. IgG antibodies to hsp90 and IgM antibodies to the 73-kD member of the hsp70 family are present in patients with SLE and, less frequently, other systemic autoimmune disorders (MINOTA et al. 1988a, b). The anti-hsp70 antibodies are restricted to the constitutively expressed 73-kD member of this family and cross-react with a peptide-binding protein in the mouse which plays a role in antigen presentation (LAKEY et al. 1987; VAN BUSKIRK et al. 1989; JARJOUR et al., unpublished results). A second member of the hsp70 family, the recently described mitochondrial protein grp75 (MIZZEN et al. 1989), also is a target of IgM autoantibodies in SLE (MINOTA and WINFIELD, unpublished results). Autoantibodies to ubiquitin have been reported in SLE as well (MULLER et al. 1988). Precise information concerning the diagnostic specificity, linkage with other autoantibody systems, relationship to disease activity status, and reactive epitopes of anti-stress protein autoantibodies is not available.

8.2 Rheumatoid Arthritis

Elevated IgG and IgA antibodies to recombinant hsp60 and stress proteins from mycobacteria have been reported in patients with rheumatoid arthritis, SLE, and Crohn's disease (IgA class only) (BAHR et al. 1988a, b; TSOULFA et al. 1988). The data concerning associations of autoantibodies to mycobacteria with MHC class II antigens are inconsistent. A preliminary report suggests that 50% of patients with rheumatoid arthritis have autoantibodies to *Borrelia burgdorferi* hsp60 (WEISS et al. 1990).

8.3 Ankylosing Spondylitis

Sera from 39% of patients with ankylosing spondylitis, but not other types of arthritis, have antibodies that stain a heat shock puff at locus 93D in polytene chromosomes in salivary glands of *Drosophila* larvae (LAKOMEK et al. 1984). This interesting observation recently was confirmed and extended to include specific staining of a second heat shock puff, 87AC (BRAND et al. 1989). The latter study also demonstrated that 49% of ankylosing spondylitis sera and 1/2 Reiter's disease sera stain a heat-inducible 63-kD *Drosophila* protein (? homologous to the hsp60 protein of mycobacteria). Antibodies of this type are not detected in rheumatoid arthritis and only rarely in SLE and normal individuals. The relevance of these interesting observations to the association of the spondyloarthropathies with HLA-B27 and gram-negative organisms (KINSELLA et al. 1983; PRENDERGAST et al. 1984; VAUGHN 1990; HUSBY et al. 1988, 1989) remains to be established.

9 Cell Surface Expression of Stress Proteins

Localization of stress proteins on plasma membranes is of great interest with respect to the issue of T cell autoreactivity, and is to be expected. Heat shock causes unfolding of membrane proteins in mammalian cells analogous to heat-induced denaturation of water-soluble proteins (LEPOCK et al. 1983), and this is accompanied by glycosylation of specific heat shock proteins (HENLE et al. 1988) which appear to be involved in membrane-stabilizing effects (SHIVERS et al. 1988). Surface membrane expression of stress proteins can be induced by virus infection of the cell (LA THANGUE and LATCHMAN 1988; KOGA et al. 1989), certain drugs (e.g., sodium valproate) (MARTIN and REGAN 1988), and certain cytokines (KOGA et al. 1989). In addition to conventional presentation by MHC class I and class II molecules of exogenous stress proteins and autologous stress proteins which have been shed or released from cells, at least four other mechanisms can be envisioned by which stress proteins could be expressed on the cell surface: (a) presentation of autologous stress protein peptides by class IB molecules, (b) translocation of stress proteins to the cell surface as they chaperone MHC or other nascent integral membrane proteins to the plasma membrane, (c) direct binding of circulating extracellular self stress proteins to surface membrane proteins, and (d) expression of stress proteins or closely related proteins on the plasma membrane as integral membrane proteins.

Evidence is beginning to accumulate in support of several of these potential mechanisms. First, the self-stimulatory activity of mycobacterial hsp60-reactive $\gamma\delta$ thymocyte hybridomas in newborn mice (O'BRIEN et al. 1989) and the killing of mouse bone marrow macrophages, in the absence of exogenous antigen, by mycobacterial hsp60-primed CTLs (KOGA et al. 1989) both suggest that autologous stress protein peptides that cross-react with mycobacterial hsp60 are recognized by T cell receptors. Second, the detection of hsp70 as a peptide-binding protein on the surface of antigen-presenting cells may reflect the chaperone function of this family of stress proteins as it accompanies MHC class II/antigen peptide complexes to the plasma membrane (LAKEY et al. 1987; VANBUSKIRK et al. 1989). Third, physiological release of stress proteins has been documented by observations that grp100 is shed by human fibroblasts (MCCORMIC et al. 1982) and that cultured rat embryo cells can be stimulated to selectively release hsp110 and several members of the hsp70 family (HIGHTOWER and GUIDON 1989). Fourth, certain proteins homologous to grp94 and hsp90 in the hsp90 family have been characterized biochemically as bona fide integral membrane proteins (MAZZARELLA and GREEN 1987; LA THANGUE and LATCHMAN 1988). Mammalian stress proteins that have been localized in cell membranes include: various members of the hsp70, hsp90, and hsp100 families (LAKEY et al. 1987; VANBUSKIRK et al. 1989; MAZZARELLA and GREEN 1987; LA THANGUE and LATCHMAN 1988; POUYSSEGUR and YAMADA 1978; SHIU et al. 1977; MCCORMIC et al. 1982; JARJOUR et al. 1989); an isoform of microtubule-associated protein (hsp70 related) in synaptosomal membranes of brain (LIM et al. 1984; WHATLEY et al.

1986); a 83-kD stress protein in normal, heat-shocked, and recovering cultured cells (CARBAJAL et al. 1986); grp78 (CHANG et al. 1987; BALDWIN et al. 1987); and, possibly, stress proteins induced by heat shock in brain capillary endothelium (SHIVERS et al. 1988). In short, considerable data suggest that many autologous stress proteins are exposed to the external environment of the cell under certain circumstances. Regardless of mechanism, this supports the concept that both uninfected cells expressing autologous stress proteins (or peptides) and infected cells expressing cross-reactive microbial stress protein stimulate T cells and/or become the targets of immunological attack, with resultant cell and tissue injury. This would explain, for example, the persistence of aggressive T cell auto-reactivity in the synovium of patients with rheumatoid arthritis or reactive arthritis long after the triggering microorganism had disappeared.

Of interest in this regard are data from the authors' laboratory which suggest that hsp60 cross-reactive protein may be constitutively expressed on human $\gamma\delta$ cells (unpublished observations). In indirect immunofluorescence, immunoblotting, and immunoprecipitation experiments, two polyclonal rabbit antisera against human hsp60 detect surface expression of a ~ 77-kD molecule on two $\gamma\delta$ T cell lines (PEER and a human thymocyte line). This molecule could not be demonstrated on peripheral blood T cells, on T cell lines having $\alpha\beta$ TCRs, or on T cells lacking TCRs altogether, and is distinct from intracellular hsp60 and known members of the hsp70 family. These observations suggest that constitutive surface expression of an hsp60-related determinant(s) may be a general property of $\gamma\delta$ lymphocytes, and may provide an alternative explain for the "self-reactivity" of neonatal $\gamma\delta$ thymocyte hybridomas (O'BRIEN et al. 1989).

10 Summary

At birth, the immune system is biased toward recognition of microbial antigens in order to protect the host from infection. Recent data suggest that an important initial line of defense in this regard involves autologous stress proteins, especially conserved peptides of hsp60, which are presented to T cells bearing $\gamma\delta$ receptors by relatively nonpolymorphic class Ib molecules. Natural antibodies may represent a parallel B cell mechanism. Through an evolving process of "physiological" autoreactivity and selection by immunodominant stress proteins common to all prokaryotes, B and T cell repertoires expand during life to meet the continuing challenge of infection. Because stress proteins of bacteria are homologous with stress proteins of the host, there exists in genetically susceptible individuals a constant risk of autoimmune disease due to failure of mechanisms for self–nonself discrimination. That stress proteins actually play a role in atuoimmune processes is supported by a growing body of evidence which, collectively, suggests that autoreactivity in chronic inflammatory arthritis involves, at least initially, $\gamma\delta$ cells which recognize epitopes of the stress protein hsp60. Alternate mechanisms for T cell stimulation by stress proteins undoubtedly also exist, e.g., molecular mimicry of the DRβ third hypervariable region susceptibility locus for rheumatoid arthritis by a DnaJ stress protein epitope in gram-negative bacteria. While there still is confusion with respect to the most relevant stress protein epitopes, a central role for stress proteins in the etiology of arthritis appears likely. Furthermore, insight derived from the work thus far in adjuvant-induced arthritis already is stimulating analyses of related phenomena in autoimmune diseases other than those involving joints. Only limited data are available in the area of humoral autoimmunity to stress proteins. Autoantibodies to a number of stress proteins have been identified in SLE and rheumatoid arthritis, but their pathogenetic significance remains to be established. Nevertheless, the capacity of certain stress proteins to bind to multiple proteins in the nucleus and cytoplasm both physiologically and during stress or injury to cells, suggests that stress proteins may be important elements in the "immunogenic particle" concept of the origin of antinuclear and other autoantibodies. In short, this fascinating group of proteins, so mysterious only a few years ago, has impelled truly extraordinary new lines of investigation into the nature of autoimmunity and autoimmune disease.

Acknowledgments. Research in the authors' laboratory is supported by National Institutes of Health grants R01 AM30863, T32 AR7416, and P60 AR30701, and a Biomedical Research Center grant from the Arthritis Foundation.

References

Aho K, Leirisalo Repo M, Repo H (1985) Reactive arthritis. Clin Rheum Dis 11: 25–40
Anderson DC, Barrey ME, Buchanan TM (1988) Exact definition of species-specific and cross-reactive epitopes of the 65-kilodalton protein of *Mycobacterium leprae* using synthetic peptides. J Immunol 141: 607–613
Arnett FC, Bias WB, Reveille JD (1989) Genetic studies in Sjögren's syndrome and systemic lupus erythematosus. J Autoimmun 2: 403–413
Atkinson JP (1988) Complement deficiency: predisposing factor to autoimmune syndromes. Am J Med 85: 45–47
Bahr GM, Rook GAW, Al-Saffar M, van Embden J, Stanford JL, Behbehani K (1988a) Antibody levels to mycobacteria in relation to HLA type: evidence for non-HLA-linked high levels of antibody to the 65 kD heat shock protein of *M. bovis* in rheumatoid arthritis. Clin Exp Immunol 74: 211–215
Bahr GM, Rook GAW, Shabin A, Stanford JL, Sattar MI, Behbehani K (1988b) HLA-DR-associated isotype-specific regulation of antibody levels to mycobacteria in rheumatoid arthritis. Clin Exp Immunol 72: 26
Baldwin GS, Chandler R, Seet KL, Weinstock J, Grego B, Rubira M, Mortiz RL, Simpson RJ (1987) Structural studies on a 75-kDa glycoprotein isolated from porcine gastric mucosal membranes: close homology with the 78-kDa glucose-regulated family of proteins. Protein Seq Data Anal 1: 7–12
Balkwill FR, Burke F (1989) The cytokine network. Immunol Today 10, No. 9: 299–304
Banerjee S, Haqqi TM, Luthra HS, Stuart JM, David CS (1988) Possible role of V beta T cell receptor genes in susceptibility to collagen-induced arthritis in mice. J Exp Med 167: 832–839

Bardwell JC, Craig EA (1984) Major heat shock gene of *Drosophila* and the *Escherichia coli* heat-inducible dnaK gene are homologus. Proc Natl Acad Sci USA 81: 848–852

Berberian PA, Myers W, Tytell M, Challa V, Bond Mg (1990) Immunohistochemical localization of heat shock protein-70 in normal-appearing and atherosclerotic specimens of human arteries. Am J Pathol 136: 71–80

Billingham MEJ, Carney S, Butler R, Colston MJ (1990) A mycobacterial 65-kD heat shock protein induces antigen-specific suppression of adjuvant arthritis, but is not itself arthritogenic. J Exp Med 171: 339–344

Bond U, Schlesinger MJ (1987) Heat-shock proteins and development. Adv Genet 24: 1–29

Born W, Happ MP, Dallas A, Reardon C, Kubo R, Shinnick T, Brennan P, O'Brien R (1990a) Recognition of heat shock proteins and gamma/delta cell function. Immunol Today 11: 40–43

Born W, Hall L, Dallas A, Boymel J, Shinnick T (1990b) Recognition of a peptide antigen by heat shock reactive $\gamma\delta$ T lymphocytes. Science 249: 67–69

Bottazzo GF, Pujol Borrell R, Hanafusa T, Feldmann M (1983) Role of aberrant HLA-DR expression and antigen presentation in induction of endocrine autoimmunity. Lancet II: 1115–1119

Boylston AW (1987) The T-cell antigen receptor gamma chain and its accomplices. Immunol Today 8: 144–145

Brand SR, McIntosh DP, Bernstein RM (1989) Antibody to a 63 kD protein in ankylosing spondylitis (abstr). Br J Rheumatol 28 [Suppl 1]: 5

Bernnan FM, Londei M, Jackson AM, Hercend T, Bernner MB, Maini RN, Feldmann M (1988) T cells expressing gamma delta chain receptors in rheumatoid arthritis. J Autoimmun 1: 319–326

Brenner MB, Strominger JL, Krangel MS (1988) The gamma delta T cell receptor. Adv Immunol 43: 133–192

Brett SJ, Lamb JR, Cox JH, Rothbard JB, Mehlert A, Ivanyi J (1989) Differential pattern of T cell recognition of the 65-kDa mycobacterial antigen following immunization with the whole protein or peptides. Eur J Immunol 19: 1303–1310

Brunet S, Giacomoni PU (1989) Heat shock mRNA in mouse epidermis after UV irradiation. Mutat Res 219: 217–224

Caccia N, Mak TW (1988) T cell receptors. Am J Med 85: 9–11

Caltabiano MM, Koestler TP, Poste G, Greig RG (1986) Induction of mammalian stress proteins by a triethylphosphine gold compound used in the therapy of rheumatoid arthritis. Biochem Biophys Res Commun 138: 1074–1080

Caltabiano MM, Poste G, Greig RG (1988) Induction of the 32-kD human stress protein by auranofin and related triethylphosphine gold analogs. Biochem pharmacol 37: 4089–4093

Cannon GW, McCall S (1989) Inhibition of the passive transfer of adjuvant-induced arthritis in rats by gold sodium thiomalate (abstr). Arthritis Rheum 32: S54–S54

Carbajal ME, Duband JL, Lettre F, Valet JP, Tanguay RM (1986) Cellular localization of *Drosophila* 83-kilodalton heat shock protein in normal, heat-shocked, and recovering cultured cells with a specific antibody. Biochem Cell Biol 64: 816–825

Chang SC, Wooden SK, Nakaki T, Kim YK, Lin AY, Kung L, Attenello JW, Lee AS (1987) Rat gene encoding the 78-kDa glucose-regulated protein GRP78: its regulatory sequences and the effect of protein glycosylation on its expression. Proc Natl Acad Sci USA 84: 680–684

Chen PP, Siminovitch KA, Olsen NJ, Erger RA, Carson DA (1989) A highly informative probe for two polymorphic Vh gene regions that contain one or more autoantibody-associated Vh genes. J Clin Invest 84: 706–710

Cherrie AH, McLean L, Archer JR (1989) Immunoblot for stress proteins in synovial fluid (abstr). Arthritis Rheum 32: S152–S152

Clerget M, Polla BS (1990) Erythrophagocytosis induce heat shock protein synthesis by human monocytes-macrophages. Proc Natl Acad Sci USA 87: 1081–1085

Cohen IR (1988) The self, the world and autoimmunity. Sci Am 258: 52–60

Cohen IR (1989a) Autoimmunity to receptors. Isr J Med Sci 25: 695–697

Cohen IR (1989) An eclectic summary of the symposium on autoimmunity. Immunol Today 10: 394–396

Cohen IR, Holoshitz J, Van Eden W, Frenkel A (1985) T lymphocyte clones illuminate pathogenesis and affect therapy of experimental arthritis. Arthritis Rheum 28: 841–845

Collins PC, Hightower LE (1982) Newcastle disease virus stimulates the cellular accumulation of stress (heat shock) mRNAs and proteins. J Virol 44: 703–707

Coutinho A (1989) Beyond clonal selection and network. Immunol Rev 110: 63–87

Coutinho A, Bandeira A (1989) Tolerize one, tolerize them all: tolerance is self-assertion. Immunol Today 10: 264–266

Deguchi Y, Negoro S, Kishimoto S (1987) Heat-shock protein synthesis by human peripheral mononuclear cells from SLE patients. Biochem Biophys Res Commun 148: 1063–1068

Deguchi Y, Negoro S, Kishimoto S (1988) Age-related changes of heat shock protein gene transcription in human peripheral blood mononuclear cells. Biochem Biophys Res Commun 157: 580–584

DeJoy SQ, Ferguson KM, Sapp TM, Sabriskie JB, Oronsky AL, Kerwar SS (1989) Streptococcal cell wall arthritis. J Exp Med 170: 369–382

De Maria A, Malnati M, Moretta A, Pende D, Bottino C, Casorati G, Cottafava F, Melioli G, Mingari MC, Migone N, Romagnani S, Moretta L (1987) CD3$^+$4$^-$8$^-$WT31$^-$ (T cell receptor gamma$^+$) cells and other unusual phenotypes are frequently detected among spontaneously interleukin 2-responsive T lymphocytes present in the joint fluid in juvenile rheumatoid arthritis. A clonal analysis. Eur J Immunol 17: 1815–1819

Dubois P (1989) Heat shock proteins and immunity. Res Immunol 140: 653–659

Dudani AK, Gupta RS (1989) Immunological characterization of a human homolog of the 65-kilodalton mycobacterial antigen. Infect Immun 57: 2786–2793

Dziarski R (1988) Autoimmunity:polyclonal activation or antigen induction? Immunol Today 9: 340–342

Elias D, Markovits D, Reshef T, van der Zee R, Cohen IR (1990) Induction and therapy of autoimmune diabetes in the non-obese diabetic (NOD/Lt) mouse by a 65-kDa heat shock protein. Proc Natl Acad Sci USA 87: 1576–1580

Ellis RJ, Hemmingsen SM (1989) Molecular chaperones: proteins essential for the biogenesis of some macromolecular structures. Trends Biochem Sci 14: 339–342

Faassen AE, OLeary JJ, Rodysill KJ, Bergh N, Hallgren HM (1989) Diminished heat-shock protein synthesis following mitogen stimulation of lymphocytes from aged donors. Exp Cell Res 183: 326–334

Farrelly FW, Finkelstein DB (1984) Complete sequence of heat shock-inducible hsp90 gene of *Saccharomyces cerevisiae*. J Biol Chem 259: 5745–5751

Feldmann M (1989) Molecular mechanisms involved in human autoimmune diseases: relevance of chronic antigen presentation. Class II expression and cytokine production. Immunol Suppl 2: 66–71

Ferrans VJ, Rodriguez ER (1985) Cardiovascular lesions in collagen-vascular diseases. Heart Vessels Suppl 1: 256–261

Ferris DK, Harel Bellan A, Morimoto RI, Welch WJ, Farrar WL (1989) Mitogen and lymphokine stimulation of heat shock proteins in T lymphocytes. Proc Natl Acad Sci USA 85: 3850–3854

Fleischer B (1989) Bacterial toxins as probes for the T-cell antigen receptor. Immunol Today 10: 262–264

Ford DK, da Roza D, Schulzer M (1985) Lymphocytes from the site of disease but not blood lymphocytes indicate the cause of arthritis. Ann Rheum Dis 44: 701–710

Friedman SM, Posnett DN, Tumang JR, Crow MK (1990) A potential role for microbial superantigens in the pathogenesis of systemic autoimmune disease. Arthritis Rheumatol (in press)

Froude J, Gibofsky A, Buskirk DR, Khanna A, Zabriskie JB (1989) Cross-reactivity between streptococcus and human tissue: a model of molecular mimicry and autoimmunity. Curr Top Microbiol Immunol 145: 5–26

Fujinami RS, Oldstone MB (1989) Molecular mimicry as a mechanism for virus-induced autoimmunity. Immunol Res 8: 3–15

Gaston JSH, Life PF, Bailey L, Bacon PA (1988) Synovial fluid T cells and 65 kD heat-shock protein. Lancet II: 856–856

Gaston JSH, Life PF, Bailey LC, Bacon PA (1989) In vitro responses to a 65-kilodalton mycobacterial protein by synovial T cells from inflammatory arthritis patients. J Immunol 143: 2494–2500

Gaston JSH, Life PF, Jenner PJ, Colston MJ, Bacon PA (1990) Recognition of a mucobacteria-specific epitope in the 65-kD heat-shock protein by synovial fluid-derived T cell clones. J Exp Med (in press)

Granelli Piperno A, Andrus L, Steinman RM (1986) Lymphokine and nonlymphokine mRNA levels in stimulated human T cells: kinetics, mitogen requirements, and effects of cyclosporin A. J Exp Med 163: 922–937

Granfors K, Jalkanen S, Von Essen R, Lahesmaa Rantala R, Isomaki O, Pekkola Heino K, Merilahti Palo R, Saario R, Isomaki H, Toivanen A (1989) *Yersinia* antigens in synovial-fluid cells from patients with reactive arthritis [see comments]. N Engl J Med 320: 216–221

Gregersen PK (1989) HLA Class II polymorphism: implications for genetic susceptibility to autoimmune disease. Lab Invest 61: 5–19

Grunebaum E, Malatzky Goshen E, Shoenfeld Y (1989) Natural killer cells and autoimmunity. Immunol Res 8: 292–304

Gupta RS (1990) Sequence and structural homology between a moust T-complex protein TCP-1 and the "chaperonin" family of bacterial (GroEL, 60–65 kDa heat shock antigen) and eukaryotic proteins. Biochem Int 24: 833–841

Haas IG, Meo T (1988) cDNA cloning of the immunoglobulin heavy chain binding protein. Proc Natl Acad Sci USA 85: 2250–2254

Haas IG, Wabl M (1983) Immunoglobulin heavy chain binding protein. Nature 306: 387–389

Haire RN, Peterson MS, O'Leary JJ (1988) Mitogen activation induces the enhanced synthesis of two heat-shock proteins in human lymphocytes. J Cell Biol 106: 883–891

Haqqi TM, Banerjee S, Jones WL, Anderson G, Behlke MA, Loh Dy, Luthra HS, David CS (1989) Identification of T-cell receptor V beta deletion mutant mouse strain AU/ssJ (H–2q) which is resistant to collagen-induced arthritis. Immunogenetics 29: 180–185

Haregewoin A, Soman G, Hom RC, Finberg RW (1989) Human gamma delta + T cells respond to mycobacterial heat-shock protein. Nature 340: 309–312

Hemmingsen SM, Woolford C, van der Vies SM, Tilly K, Dennis DT, Georgopoulos CP, Hendrix RW, Ellis RJ (1988) Homologous plant and bacterial proteins chaperone oligomeric protein assembly. Nature 333: 330–334

Henle KJ, Nagle WA, Norris JS, Moss AJ (1988) Enhanced glycosylation of a 50 kD protein during development of thermotolerance in CHO cells. Int J Radiat Biol 53: 839–847

Herzenberg LA, Lalor PA, Stall AM (1989) Are Ly-1 B cells important in autoimmune disease. J Autoimmun 2: 225–231

Hightower LE, Guidon PT Jr (1989) Selective release from cultured mammalian cells of heat-shock (stress) proteins that resemble glia-axon transfer proteins. J Cell Physiol 138: 257–266

Holmadhl R, Andersson ME, Goldschmidt TJ, Jansson L, Karlsson M, Malmstrom V, Mo J (1989) Collagen induced arthritis as an experimental model for rheumatoid arthritis. Immunogenetics, pathogenesis and autoimmunity. APIMS 97: 575–584

Holoshitz Z, Drucker I, Yaretzky A, Van Eden W, Klajman A, Lapidot Z, Frenkel A, Cohen IR (1986) T lymphocytes of rheumatoid arthritis patients show augmented reactivity to a fraction of mycobacteria cross-reactive with cartilage. Lancet 2: 305–309

Holoshitz J, Koning F, Coligan JE, de Bruyn J, Strober S (1989) Isolation of CD4- CD8- mycobacteria-reactive T lymphocyte clones from rheumatoid arthritis synovial fluid. Nature 339: 226–229

Hough AJ Jr, Rank RG (1989) Pathogenesis of acute arthritis due to viable Chlamydia trachomat. Am J Pathol 134: 903–912

Hurst NP (1990) Stress (heat shock) proteins and rheumatic disease, new advance or just another band wagon? Rheumatol Int 9: 271–276

Husby G, Tsuchiya N, Schwimmbeck PL, Keat A, Pahle JA, Oldstone MB, Williams RC Jr (1988) Synovial expression of HLA-B27 related antigens in ankylosing spondylitis. Scand J Rheumatol [Suppl] 76: 23–25

Husby G, Tsuchiya N, Schwimmbeck PL, Keat A, Pahle JA, Oldstone MB, Williams RC Jr (1989) Cross-reactive epitope with Klebsiella pneumoniae nitrogenase in articular tissue of HLA-B27 + patients with ankylosing spondylitis. Arthritis Rheum 32: 437–445

Inoue K (1989) C5 neoepitopes appearing during activation. Complement Inflamm 6: 219–222

Janeway CA Jr, Fischer-Lindahl K, Hammerling U (1988) The Mls locus: new clues to a lingering mystery. Immunol Today 9, No. 5: 125–126

Janeway CA Jr, Yagi J, Conrad PJ, Katz ME, Jones B, Buxser S (1989) T-cell responses to Mls and to bacterial proteins that mimic its behavior. Immunol Rev 107: 61–88

Janis EM, Kaufmann HE, Schwartz RH, Pardoll DM (1989) Activation of $\gamma\delta$ T cells in the primary immune response to Mycobacterium tuberculosis. Science 244: 713–716

Jarjour W, Tsai V, Woods V, Welch W, Pierce S, Shaw M, Mehta H, Dillmann W, Zvaifler N, Winfield J (1989) Cell surface expression of heat shock proteins. Arthritis Rheum 32: S44

Jay G, Shiu RP, Jay FT, Levine AS, Pastan I (1978) Identification of a transformation-specific protein induced by a Rous sarcoma virus. Cell 13: 527–534

Jindal S, Dudani AK, Singh B, Harley CB, Gupta RS (1989) Primary structure of a human mitochondrial protein homologous to the bacterial and plant chaperonins and to the 65-kilodalton mycobacterial antigen. Mol Cell Biol 9: 2279–2283

Kabelitz D, Bender A, Schondelmaier S, Schoel B, Kaufmann SHE (1990) A large fraction of human peripheral blood y/b + T cells is activated by Mycobacterium tuberculosis but not by its 65-kD heat shock protein. J Exp Med 171: 667–679

Karlsson-Parra A, Söderström K, Ferm M, Ivanyi J, Kiessling R, Klareskog L (1990) Presence of human 65 kD heat shock protein (hsp) in inflamed joints and subcutaneous nodules of RA patients. Scand J Immunol 31: 283–288

Karounos DG, Wolinsky JS, Thomas JW (1990) Autoantibodies to 52 kDa islet proteins: a disease and tissue-specific marker for insulin-dependent diabetes mellitus (abstr). Clin Res 38: 408A

Kaufmann SHE (1990) Heat shock proteins and the immune response. Immunol Today 11: 129–136

Kaufmann SH, Flesch IE, Munk ME, Wand Wurttenberger A, Schoel B, Koga T (1989) Cell-mediated immunity to mycobacteria: a double-sided sword. Rheumatol Int 9: 181–186

Keat A, Thomas B, Dixey J, Osborn M, Sonnex C, Taylor Robinson D (1987) Chlamydia trachomatis and reactive arthritis: the missing link. Lancet I: 72–74

Keat A, Thomas B, Hughes R, Taylor Robinson D (1989) Chlamydia trachomatis in reactive arthritis. Rheumatol Int 9: 197–200

Kinsella TD, Fritzler MJ, McNeil DJ (1983) Ankylosing spondylitis. A disease in search of microbes. J Rheumatol 10: 2–4

Kofler R, Dixon FJ, Theofilopoulos AN (1989) Genetic basis for autoantibody-production in murine models of systemic autoimmunity. Contrib Microbiol Immunol 11: 206–230

Koga T, Wand Wurttenberger A, DeBruyn J, Munk ME, Schoel B, Kaufmann SH (1989) T cells against a bacterial heat shock protein recognize stressed macrophages. Science 245: 1112–1115

Köller R, Brom C, Brom J, König W (1989) Heat shock induces alterations of lipoxygenase pathway in human polymorphonuclear granulocytes. Prostaglandins Leukortrienes and Essential Fatty Acids 38: 99–106

Kroemer G, Wick G (1989) The role of interleukin 2 in autoimmunity. Immunol Today 10: 246–251

Kubo T, Towle CA, Mankin HJ, Treadwell BV (1985) Stress-induced proteins in chondrocytes from patients with osteoarthritis. Arthritis Rheum 28: 1140–1145

Kumkumian GK, Lafyatis R, Remmers EF, Case JP, Kim SJ, Wilder RL (1989) Platelet-derived growth factor and IL-1 interactions in rheumatoid arthritis. Regulation of synoviocyte proliferation, prostaglandin production, and collagenase transcription. J Immunol 143: 833–837

Lafaille JJ, Haas W, Coutinho A, Tonegawa S (1990) Positive selection of gamma delta T cells. Immunol Today 11, No. 3: 75–78

Lafyatis R, Thompson NL, Remmers EF, Flanders KC, Roche NS, Kim SJ, Case JP, Sporn MB, Roberts AB, Wilder RL (1989) Transforming growth factor-beta production by synovial tissues from rheumatoid patients and streptococcal cell wall arthritic rats. Studies on secretion by synovial fibroblast-like cells and immunohistologic localization. J Immunol 143: 1142–1148

Lakey EK, Margoliash E, Pierce SK (1987) Identification of a peptide binding protein that plays a role in antigen presentation. Proc Natl Acad Sci USA 84: 1659–1663

Lakomek HJ, Will H, Zech M, Kruskemper HL (1984) A new serologic marker in ankylosing spondylitis. Arthritis Rheum 27: 961–967

Lamb JR, Lathigra R, Rothbard JB, Sweetser D, Young RA, Ivanyi J, Young DB (1989) Identification of mycobacterial antigens recognized by T lymphocytes. Rev Infect Dis 11: S443–S447

Landry SJ, Bartlett SG (1989) The small subunit of ribulose-1,5-bisphosphate carboxylase/ oxygenase and its precursor expressed in Escherichia coli are associated with groEL protein. J Biol Chem 264: 9090–9093

Latchman DS, Norton PM, Bansal GS, Isenberg DA (1990) Function and expression of the 90 kD heat shock protein. In: Rice-Evans Stress proteins in inflammation. Richlieu, London, pp

La Thangue NB, Latchman DS (1988) A cellular protein related to heat-shock protein 90 accumulates during herpes simplex virus infection and is overexpressed in transformed cells. Exp Cell Res 178: 169–179

Lathigra RB, Young DB, Sweetser D, Young RA (1988) A gene from Mycobacterium tuberculosis which is homologous to the DnaJ heat shock protein of E. coli. Nucleic Acids Res 16: 1636–1636

Lecker S, Lill R, Ziegelhoffer T, Georgopoulos C, Bassford PJ Jr, Kumamoto CA, Wickner W (1989) Three pure chaperone proteins of Escherichia coli—SecB, trigger factor, and GroEL—form soluble complexes with precursor proteins in vitro. EMBO J 8: 2703–2709

Lepock JR, Cheng KH, Al Qysi H, Kruuv J (1983) Thermotropic lipid and protein transitions in Chinese hamster lung cell membranes: relationship to hyperthermic cell killing. Can J Biochem Cell Biol 61: 421–427

Levine JD, Goetzl EJ, Basbaum AI (1987) Contribution of the nervous system to the pathophysiology of rheumatoid arthritis and other polyarthritides. Rheum Dis Clin North Am 13: 369–383

Life PF, Viner NJ, Bacon PA, Gaston JSH (1990) Synovial fluid antigen-presenting cells unmask peripheral blood T cell responses to bacterial antigens in inflammatory arthritis. Clin Exp Immunol 79: 189–194

Lim L, Hall C, Leung T, Whatley S (1984) The relationship of the rat brain 68 kDa microtubule-associated protein with synaptosomal plasma membranes and with the Drosophila 70 kDa heat-shock protein. Biochem J 224: 677–680

Lindquist S (1986) The heat-shock response. Annu Rev Biochem 55: 1151–1191

Lipsky PE (1989) The control of antibody production by immunomodulatory molecules. Arthritis Rheum 32: 1345–1355

Lipsky PE, Davis LS, Cush JJ, Oppenheimer Marks N (1989) The role of cytokines in the pathogenesis of rheumatoid arthritis. Springer Semin Immunopathol 11: 123–162

Liu AY, Lin Z, Choi HS, Sorhage F, Li B (1989a) Attenuated induction of heat shock gene expression in aging diploid fibroblasts. J Biol Chem 264: 12037–12045

Liu AY, Bae Lee MS, Choi HS, Li BS (1989) Heat shock induction of HSP 89 is regulated in cellular aging. Biochem Biophys Res Commun 162: 1302–1310

Maher PA, Pasquale EB (1989) Heat shock induces protein tyrosine phosphorylation in cultured cells. J Cell Biol 108: 2029–2035

Maisch B (1989a) Autoreactive mechanisms in infective endocarditis. Springer Semin Immunopathol 11: 439–456

Maisch B (1989b) Autoreactivity to the cardiac myocyte, connective tissue and the extracellular matrix in heart disease and postcardiac injury. Springer Semin Immunopathol 11: 369–395

Maisch B (1989c) Retrospective and perspectives in the immunology of cardiac diseases. Springer Semin Immunopathol 11: 479–482

Mapp PI, Cherrie AH, Blake DR, Archer JR, Winrow VR, McLean L (1989) 70 kDa stress protein in the arthritic synovial membrane. Arthritis Rheum 32: S152

Martin ML, Regan CM (1988) The anticonvulsant sodium valproate specifically induces the expression of a rat glial heat shock protein which is identified as the collagen type IV receptor. Brain Res 459: 131–137

Matis LA, Glimcher LH, Paul WE, Schwartz RH (1983) Magnitude of response of histocompatibility-restricted T-cell clones is a function of the product of the concentrations of antigen and Ia molecules. Proc Natl Acad Sci USA 80: 6019–6023

Mattei D, Scherf A, Bensaude O, da Silva LP (1989) A heat shock-like protein from the human malaria parasite Plasmodium falciparum induces autoantibodies. Eur J Immunol 19: 1823–1828

Mazzarella RA, Green M (1987) ERp99, an abundant, conserved glycoprotein of the endoplasmic reticulum, is homologous to the 90-kDa heat shock protein (hsp90) and the 94-kDa glucose regulated protein (GRP94). J Biol Chem 262: 8875–8883

McCormic PJ, Millis AJT, Babiarz B (1982) Distribution of a 100 k dalton glucose-regulated cell surface protein in mammalian cell cultures and sectioned tissues. Exp Cell Res 138: 63–72

McLean L, Winrow VR, Mapp RI, Cherrie(AH, Archer JR, Blake DR (1988) Synovial fluid T cells and 65 kD heat-shock protein. Lancet II: 856–857

McMullin TW, Hallberg RL (1988) A highly evolutionary conserved mitochondrial protein is structurally related to the protein encoded by the Escherichia coli groEL gene. Mol Cell Biol 8: 371–380

Medof ME, Iida K, Mold C, Nussenzweig V (1982) Unique role of the complement receptor CR1 in the degradation of C3b associated with immune complexes. J Exp Med 156: 1739–1754

Mehra V, Sweetser D, Young RA (1986) Efficient mapping of protein antigenic determinants. Proc Natl Acad Sci USA 83: 7013–7017

Melancon Kaplan J, Hunter SW, McNeil M et al (1988) Immunological significance of Mycobacterium leprae cell walls. Proc Natl Acad Sci USA 85: 1917–1921

Midtvedt T(1987) Intestinal bacteria and rheumatic disease. Scand J Rheumatol 64: 49–54

Milarski KL, Welch WJ, Morimoto RI (1989) Cell cycle-dependent association of HSP70 with specific cellular proteins. J Cell Biol 108: 413–423

Miller JF, Morahan G, Allison J (1989) Immunological tolerance: new approaches using transgenic mice. Immunol Today 10: 53–57

Minota S, Cameron B, Welch WJ, Winfield JB (1988a) Autoantibodies to the constitutive 73-kD member of the hsp70 family of heat shock proteins in systemic lupus erythematosus. J Exp Med 168: 1475–1480

Minota S, Koyasu S, Yahara I, Winfield JB (1988b) Autoantibodies to the heat-shock protein hsp90 in systemic lupus erythematosus. J Clin Invest 81: 106–109

Mizisin AP, Wiley CA, Hughes RAC, Powell HC (1987) Peripheral nerve demyelination in rabbits after inoculation with Freund's complete adjuvant alone or in combination with lipid haptens. J Neuroimmunol 16: 381–395

Mizzen LA, Chang C, Garrels JI, Welch WJ (1989) Identification, characterization, and purification of two mammalian stress proteins in mitochondria, grp 75, a member of the hsp 70 family and hsp 58, a homolog of the bacterial groEL protein. J Biol Chem 264: 20664–20675

Modlin RL, Pirmez C, Hofman FM, Torigian V, Uyemura K, Rea TH, Bloom BR, Brenner MB (1989) Lymphocytes bearing antigen-specific gamma delta T-cell receptors accumulate in human infectious disease lesions. Nature 339: 544–548

Morimoto C, Romain PL, Fox DA, Anderson P, DiMaggio M, Levine H, Schlossman SF (1988) Abnormalities in CD4+ T-lymphocyte subsets in inflammatory rheumatic diseases. Am J Med 84: 817–825

Morrison RP, Belland RJ, Lyng K, Caldwell HD (1989) Chlamydial disease pathogenesis. The 57-kD chlamydial hypersensitivity antigen is a stress response protein. J Exp Med 170: 1271–1283

Muller S, Briand JP, Van Regenmortel MH (1988) Presence of antibodies to ubiquitin during the autoimmune response associated with the systemic lupus erythematosus. Proc Natl Acad Sci USA 85: 8176–8180

Munk ME, Schoel B, Kaufmann SH (1988) T cell responses of normal individuals towards recombinant protein antigens of Mycobacterium tuberculosis. Eur J Immunol 18: 1835–1838

Munk ME, Schoel B, Modrow S, Karr RW, Young RA, Kaufmann SH (1989) T lymphocytes from healthy individuals with specificity to self-epitopes shared by the mycobacterial and human 65-kilodalton heat shock protein. J Immuno 143: 2844–2849

Munro S, Pelham HR (1986) An Hsp70-like protein in the ER: identity with the 78 kd glucose-regulated protein and immunoglobulin heavy chain binding protein. Cell 46: 291–300

Mustafa AS, Oftung F, Deggerdal A, Gill HK, Young RA, Godal T (1988) Gene isolation with human T lymphocyte probes. Isolation of a gene that expresses an epitope recognized by T cells specific for Mycobacterium bovis BCG and pathogenic mycobacteria. J Immunol 141: 2729–2733

Nepom GT (1989a) Structural variation among major histocompatibility complex class-II genes which predispose to autoimmunity. Immunol Res 8: 16–38

Nepom GT (1989b) Determinants of genetic susceptibility in HLA-associated autoimmune disease. Clin Immunol Immunopathol 53: S53–S62

Nepom GT, Byers P, Seyfried C, Healey LA, Wilske KR, Stage D, Nepom BS (1989) HLA genes associated with rheumatoid arthritis. Identification of susceptibility alleles using specific oligonucleotide probes. Arthritis Rheum 32: 15–21

Nevins JR (1982) Induction of the synthesis of a 70,000 dalton mammalian heat shock protein by the adenovirus E1A gene product. Cell 29: 913–919

Nielsen SL, Peter JB (1990) B. burgdorferi persistently detected by polymerase chain reaction in synovial fluid of a patient with lyme arthritis resistant to therapy (abstr). Clin Res 38: 402A–402A

Norton PM, Isenberg DA, Latchman DS (1989) Elevated levels of the 90 kd heat shock protein in a proportion of SLE patients with active disease. J Autoimmun 2: 187–195

Oblas B, Boyd ND, Luber-Narod J, Reyes VE, Leeman SE (1990) Isolation and identification of a polypeptide in the HSP 70 family that binds substance P. Biochem Biophys Res Commun 166: 978–983

O'Brien RL, Happ MP, Dallas A, Palmer E, Kubo R, Born WK (1989) Stimulation of a major subset of lymphocytes expressing T cell receptor γδ by an antigen derived from Mycobacterium tuberculosis. Cell 57: 667–674

Oftung F, Mustafa AS, Shinnick TM, Houghten RA, Kvalheim G, Degre M, Lundin EA, Godal T (1988) Epitopes of the Mycobacterium tuberculosis 65-kilodalton protein antigen as recognized by human T cells. J Immunol 414: 2749–2754

Oldstone MBA (1987) Molecular mimicry and autoimmune disease. Cell 50: 819–820

Oldstone MB (1989) Virus-induced autoimmunity: molecular mimicry as a route to autoimmune disease. J Autoimmun 2: 187–194

Opperman H, Levinson W, Bishop JM (1981) A cellular protein that associated with a transforming protein of Rous sarcoma virus is also a heat-shock protein. Proc Natl Acad Sci USA 78: 1067–1071

Ottenhoff TH, Torres P, de las Aguas JT, Fernandez R, Van Eden W, De Vries RR, Stanford JL (1986) Evidence for an HLA-DR4-associated immune-response gene for Mycobacterium tuberculosis. A clue to the pathogenesis of rheumatoid arthritis. Lancet II: 310–313

Ottenhoff TH, Ab BK, van Embden JD, Thole JE, Kiessling R (1988) The recombinant 65-kD heat shock protein of Mycobacterium bovis Bacillus Calmette-Guerin/M. tuberculosis is a target molecule for CD4+ cytotoxic T lymphocytes that lyse human monocytes. J Exp Med 168: 1947–1952

Pallas DC, Morgan W, Roberts TM (1989) The cellular proteins which can associate specifically with polymavirus middle T antigen in human 293 cells include the major human 70-kilodalton heat shock proteins. J Virol 63: 4533–4539

Parham P (1989) MHC molecules. A profitable lesson in heresy [news]. Nature 340: 426–428

Payan DG (1989) Neuropeptides and inflammation: the role of substance P. Annu Rev Med 40: 341–352

Pereira P, Bandeira A, Coutinho A, Marcos MA, Toribio M, Martinez C (1989) V-region connectivity in T cell repertoires. Annu Rev Immunol 7: 209–249

Phillips PE (1988) The role of infectious agents in the spondylarthropathies. Scand J Rheumatol 17: 435–443

Polla BS (1988) A role for heat shock proteins in inflammation. Immunol Today 9: 134–137

Porcelli S, Brenner MB, Greenstein JL, Balk SP, Terhorst C, Bleicher PA (1989) Recognition of cluster of differentiation 1 antigens by human $CD4^-CD8^-$ cytolytic T lymphocytes. Nature 341: 447–450

Pouyssegur J, Yamada KM (1978) Isolation and immunological characterization of a glucose-regulated fibroblast cell surface glycoprotein and its nonglycosylated precursor. Cell 13: 139–150

Prendergast JK, Sullivan JS, Geczy AF, Upfold LI, Edmonds JP, Bashir HV (1984) The enigma of the Klebsiella connection and ankylosing spondylitis: a commentary. Hum Immunol 9: 131–136

Rajasekar R, Sim G-K, Augustin A (1990) Self heat shock and $\gamma\delta$ T-cell reactivity. Proc Natl Acad Sci USA 87: 1767–1771

Reading DS, Hallberg RL, Myers AM (1989) Characterization of the yeast HSP60 gene coding for a mitochondrial assembly factor. Nature 337: 655–659

Rebbe NF, Hickman WS, Ley TJ, Stafford DW, Hickman S (1989) Nucleotide sequence and regulation of a human 90-kDa heat shock protein gene. J Biol Chem 264: 15006–15011

Renkonen R, Ristimaki A, Hayry P (1988) Interferon-gamma protects human endothelial cells from lymphokine-activated killer cell-mediated lysis. Eur J Immunol 18: 1839–1842

Requena JM, Lopez MC, Jimenez Ruiz A, Morales G, Alonso C (1989) Complete nucleotide sequence of the hsp70 gene of T. cruzi. Nucleic Acids Res 17: 797–797

Res PC, Breedveld FC, Van Embden JDA, Schaar CG, Van Eden W, Cohen IR, De Vires RRP (1988) Synovial fluid T cell reactivity against 65 kD heat shock protein of mycobacteria in early rheumatoid arthritis. Lancet II: 478–480

Richman SJ, Vedvick TS, Reese RT (1989) Peptide mapping of conformational epitopes in a human malarial parasite heat shock protein. J Immunol 143: 285–292

Ritchlin CT, Winchester RJ (1989) Potential mechanisms for coordinate gene activation in the rheumatoid synoviocyte: implications and hypotheses. Springer Semin Immunopathol 11: 219–234

Rittner C, Schneider RM (1989) Complexity of MHC class III genes and complement polymorphism. Immunol Today 10, No. 12: 401–403

Robaye B, Hepburn A, Lecocq R, Fiers W, Boeynaems JM, Dumont JE (1989) Tumor necrosis factor-alpha induces the phosphorylation of 28 kDa stress proteins in endothelial cells: possible role in protection against cytoxicity. Biochem Biophys Res Commun 163: 301–308

Romano JW, Seldin MF, Appella E (1989) Linkage of the mouse Hsp84 heat shock protein structural gene to the H-2 complex. Immunogenetics 29: 142–144

Rook GA (1988) Rheumatoid arthritis, mycobacterial antigens and agalactosyl IgG. Scand J Immunol 28: 487–493

Rose NR (1989) Pathogenic mechanisms in autoimmune diseases. Clin Immunol Immunopathol 53: S7–16

Rowe IF, Deans AC, Keat AC (1987) Pyogenic infection and rheumatoid arthritis. Postgrad Med J 63: 19–22

Saag MS, Bennett JC (1987) The infectious etiology of chronic rheumatic diseases. Semin Arthritis Rheum 17: 1–23

Sadler I, Chiang A, Kurihara T, Rothblatt J, Way J, Silver P (1989) A yeast gene important for protein assembly into the endoplasmic reticulum and the nucleus has homology to DnaJ, an Escherichia coli heat shock protein. J Cell Biol 109: 2665–2675

Sargent CA, Dunham I, Trowsdale J, Campbell RD (1989) Human major histocompatibility complex contains genes for the major heat shock protein HSP70. Proc Natl Acad Sci USA 86: 1968–1972

Sarvetnick N, Liggitt D, Pitts SL, Hansen SE, Stewart TA (1988) Insulin-dependent diabetes mellitus induced in transgenic mice by ectopic expression of class II MHC and interferon-gamma. Cell 52: 773–782

Sawai ET, Butel JS (1989) Association of a cellular heat shock protein with simian virus 40 large T antigen in transformed cells. J Viron 63: 3961–3973

Schild H, Rötzschke O, Kalbacher H, Rammeneee H-G (1990) Limit of T cell tolerance to self proteins by peptide presentation. Science 247: 1587–1589

Schlesier M, Haas G, Wolff-Vorbeck G, Melchers I, Peter H-H (1988) Autoreactive T cells in rheumatic disease (1) analysis of growth frequences and autoreactivity of T cells in patients with rheumatoid arthritis and lyme disease. J Autoimmun 2: 31–49

Schlesinger MJ (1986) Heat shock proteins: the search for functions. J Cell Biol 103: 321–325

Schuh S, Yonemoto W, Brugge J, Bauer VJ, Riehl RM, Sullivan WP, Toft DO (1985) A 90,000-dalton binding protein common to both steroid receptors and the Rous sarcoma virus transforming protein, pp60^{v-src}. J Biol Chem 260: 14292–14296

Schwartz RH (1989) Acquisition of immunologic self-tolerance. Cell 57: 1073–1081

Sheshberadaran H, Norrby E (1984) Three monoclonal antibodies against measles virus F protein cross-react with cellular stress proteins. J Virol 52: 995–999

Shinnick TM, Vodkin MH, Williams JC (1988) The *Mycobacterium tuberculosis* 65-kilodalton antigen is a heat shock protein which corresponds to common antigen and to the *Escherichia coli* GroEL protein. Infect Immun 56: 446–451

Shiu RP, Pouyssegur J, Pastan I (1977) Glucose depletion accounts for the induction of two transformation-sensitive membrane proteins in Rous sarcoma virus-transformed chick embryo fibroblasts. Proc Natl Acad Sci USA 74: 3840–3844

Shivakumar S, Tsokos GC, Datta SK (1989) T cell receptor alpha/beta expressing double-negative (CD4$^-$/CD8$^-$) and CD4$^+$ T helper cells in humans augment the production of pathogenic anti-DNA autoantibodies associated with lupus nephritis. J Immunol 143: 103–112

Shivers RR, Pollock M, Bowman PD, Atkinson BG (1988) The effect of heat shock on primary cultures of brain capillary endothelium: inhibition of assembly of zonulae occludentes and the synthesis of heat-shock proteins. Eur J Cell Biol 46: 181–195

Shoenfeld Y, Isenberg DA (1988) Mycobacteria and autoimmunity. Immunol Today 9: 178–181

Shoenfeld Y, Isenberg DA (1989) The mosaic of autoimmunity. Immunol Today 10: 123–126

Shoenfeld Y, Teplizki HA, Mendlovic S, Blank M, Mozes E, Isenberg DA (1989) The role of the human anti-DNA idiotype 16/6 in autoimmunity. Clin Immunol Immunopathol 51: 313–325

Siegelman M, Bond MW, Gallatin WM, St. John T, Smith HT, Fried VA, Weissman IL (1986) Cell surface molecules associated with lymphocyte homing is a ubiquitinated branched-chain glycoprotein. Science 231: 823–829

Sigal LH, Steere AC, Freeman DH, Dwyer JM (1986) Proliferative responses of mononuclear cells in Lyme disease: reactivity to *Borrelia burgdorferi* antigens is greater in joint fluid than in blood. Arthritis Rheum 29: 761–769

Sinclair NR, Panoskaltsis A (1988) The immunoregulatory apparatus and autoimmunity. Immunol Today 9: 260–265

Slor H, Shafrir S, Isenberg DA, Granados J, Alarcon Segovia D, Shoenfeld Y (1989) The genetics of autoimmunity: the familial tendency of systemic lupus erythematosus. Isr J Med Sci 25: 678–682

Steinhoff U, Kaufmann SHE (1988) Specific lysis by CD8$^-$ T cells of Schwann cells expressing *Mycobacterium leprae* antigens. Eur J Immunol 18: 969–972

Strahler JR, Kuick R, Eckerskorn C, Lottspeich F, Richardson BC, Fox DA, Stoolman LM, Hanson CA, Nichols D, Tueche HJ, Hanash SM (1990) Identification of two related markers for common acute lymphoblastic leukemia as heat shock proteins. J Clin Invest 85: 200–207

Strominger JL (1989) The gamma delta T cell receptor and class Ib MHC-related proteins: enigmatic molecules of immune recognition. Cell 57: 895–898

Talal N (1989) Autoimmunity and sex revisited. Clin Immunol Immunopathol 53: 355–357

Tan EM, Chan EK, Sullivan KF, Rubin RL (1988) Antinuclear antibodies (ANAs): diagnostically specific immune markers and clues toward the understanding of systemic autoimmunity. Clin Immunol Immunopathol 47: 121–141

Ternynck T, Avrameas S (1986) Murine natural monoclonal autoantibodies: a study of their polyspecificities and their affinities. Immunol Rev 94: 99–112

Theofilopoulos AN, Kofler R (1989) Molecular aspects of autoimmunity. Immunol Today 10: 180–183

Thole JR, Van Schooten WCA, Keulen WJ, Hermans PWM, Janson AAM, De Vries RRP, Kolk AHJ, Van Embden JDA (1988) Use of recombinant antigens expressed in *Escherichia coli* K-12 to map B-cell and T-cell epitopes on the immunodominant 65-kilodalton protein of *Mycobacterium bovis* BCG. Infect Immun 56: 1633–1640

Thompson SJ, Bedwell A, Hooper DC, Burtles SS, Rook GAW, Van Embden JDA, Elson CJ (1990) Pristane-induced arthritis: a possible role for the mycobacterial 65kD heat shock protein. In: Rice-Evans Stress proteins in inflammation. Richlieu, London pp

Ting LP, Tu CL, Chou CK (1989) Insulin-induced expression of human heat-shock protein gene hsp70. J Biol Chem 264: 3404–3408

Todd JA, Orbea HA, Bell JI, Chao N, Fronek Z, Jacob CO, McDermott M, Sinha AA, Timmerman L, Steinman L, McDevitt HO (1988) A molecular basis for MHC class II-associated autoimmunity. Science 240: 1003–1009

Toivanen A, Granfors K, Lahesmaa Rantala R, Toivanen P (1989) Immunological and bacteriological aspects of reactive arthritis. Rheumatol Int 9: 201–203

Tomasovic SP, Simonette RA, Wolf DA, Kelley KL, Updyke TV (1989) Co-isolation of heat stress and cytoskeletal proteins with plasma membrane proteins. Int J Hyperthermia 5: 173–190

Townsend A, Ohlen C, Bastin J, Ljunggren H-G, Foster L, Karre K (1989) Association of class I major histocompatibility heavy and light chains induced by viral peptides. Nature 340: 443–448

Tron·F, Bach JF (1989) Molecular and genetic characteristics of pathogenic autoantibodies. J Autoimmun 2: 311–320

Tsoulfa G, Rook GA, van Embden JD, Young DB, Mehlert A, Isenberg DA, Hay FC, Lydyard PM (1988) Raised serum IgG and IgA antibodies to mycobacterial antigens in rheumatoid arthritis. Ann Rheum Dis 48: 118–123

Ullrich SJ, Robinson EA, Law LW, Willingham M, Appella E (1986) A mouse tumor-specific transplantation antigen is a heat shock-related protein. Proc Natl Acad Sci USA 83: 3121–3125

Van Buskirk A, Crump BL, Margoliash, Pierce SK (1989) A peptide binding protein having a role in antigen presentation is a member of the HSP70 heat shock family. J Exp Med 170: 1799–1899

van den Broek MF, Hogervorst EJ, Van Bruggen MC, Van Eden W, van der Zee R, van den Berg WB (1989) Protection against streptococcal cell wall-induced arthritis by pretreatment with the 65-kD mycobacterial heat shock protein. J Exp Med 170: 449–466

Van Eden, De Vries RP (1989) The immunopathogenesis and immunogenetics of rheumatoid arthritis. Br J Rheumatol 28: 51–53.

Van Eden W, Holoshitz J, Nevo Z, Frenkel A, Klajman A, Cohen IR (1985) Arthritis induced by a T-lymphocyte clone that responds to *Mycobacterium tuberculosis* and to cartilage proteoglycans. Proc Natl Acad Sci USA 82: 5117–5120

Van Eden W, Holoshitz J, Cohen I (1987) Antigenic mimicry between mycobacteria and cartilage proteoglycans: the model of adjuvant arthritis. Concepts Immunopathol 4: 144–170

Van Eden W, Thole JE, van der Zee R, Noordzij A, van Embden JD, Hensen EJ, Cohen IR (1988) Cloning of the mycobacterial epitope recognized by T lymphocytes in adjuvant arthritis. Nature 331: 171–173

Van Eden W, De Vries RR (1989) The immunopathogenesis and immunogenetics of rheumatoid arthritis. Br J Rheumatol 28: 51–53

Van Eden W, Hogervorst EJ, van der Zee R, van Embden JD, Hensen EJ, Cohen IR (1989a) The mycobacterial 65 kD heat-shock protein and autoimmune arthritis Rheumatol Int 9: 187–191

Van Eden W, Hogervorst EJ, Hensen EJ, van der Zee R, van Embden JD, Cohen IR (1989b) A cartilage-mimicking T-cell epitope on a 65K mycobacterial heat-shock protein: adjuvant arthritis as a model for human rheumatoid arthritis. Curr Top Microbiol Immunol 145: 27–43

van Rooijen N (1989) Are bacterial endotoxins involved in autoimmunity by CD5+ (LY-1+) B cells. Immunol Today 10: 334–336

VanBuskirk A, Crump BL, Margoliash E, Pierce SK (1989) A peptide binding protein having a role in antigen presentation is a member of the HSP70 heat shock family. J Exp Med 170: 1799–1809

Vaughn JH (1990) Infection and rheumatic diseases: a review. Bull Rheum Dis 39: 1–7

Weiss NL, Philips MR, Sadock VA, Sigal LH, Merryman PF, Abramson SB (1990) False positive seroreactivity to *Borrelia burgdorferi* in rheumatic disease: the value of immunoblot analysis (abstr). Clin Res 38: 277A

Welch WJ, Suhan JP (1986) Cellular and biochemical events in mammalian cells during and after recovery from physiological stress. J Cell Biol 103: 2035–2052

Welch WJ, Feramisco JR, Blose SH (1985) The mammalian stress response and the cytosekelton: alterations in intermediate filaments. Ann NY Acad Sci 455: 57–67

Welch WJ, Mizzen LA, Arrigo AP (1989) Stress-induced proteins. UCLA Symposia Mol Cell Biol 96: 187–202

Whatley SA, Leung T, Hall C, Lim L (1986) The brain 68-kilodalton microtubule-associated protein is a cognate form of the 70-kilodalton mammalian heat-shock protein and is present as a specific isoform in synaptosomal membranes. J Neurochem 47: 1576–1583

White E, Spector D, Welch W (1988) Differential distribution of the adenovirus EIA proteins and colocalization of EIA with the 70-kilodalton cellular heat shock protein in infected cells. J Virol 62: 4153–4166

White J, Herman A, Pullen AM, Kubo R, Kappler JW, Marrack P (1989) The Vβ-specific superantigen staphylococcal enterotoxin B: stimulation of mature T cells and clonal deletion in neonatal mice. Cell 56: 27–35

Wilder RL, Lafyatis R, Yocum DE, Case JP, Kumkumian GK, Remmers EF (1989) Mechanisms of bone and cartilage destruction in rheumatoid arthritis: lessons from the streptococcal cell wall arthritis model in LEW/N rats. Clin Exp Rheumatol 7: S123–S127

Wilson RW, O'Brien WE, Beaudet AL (1989) Necleotide sequence of the cDNA from the mouse leukocyte adhesion protein CD18. Nucleic Acids Res 17: 5397–5397

Winchester Rj, Nonez-Roldan A (1982) Some genetic aspects of systemic lupus erythematosus. Arthritis Rheum 25: 833–837

Winfield JB (1989) Stress proteins, arthritis, and autoimmunity. Arthritis Rheum 32: 1497–1504

Winrow VR, McLean L, Morris CJ, Blake DR (1990) The heat shock protein response and its role in inflammatory disease. Ann Rheum Dis 49: 128–132

Wurst W, Benesch C, Drabent B, Rothermel E, Benecke BJ, Gunther E (1989) Localization of heat shock protein 70 genes inside the rat major histocompatibility complex close to class III genes. Immunogenetics 30: 46–49

Yamazaki M, Akaogi K, Miwa T, Imai T, Soeda E, Yokoyama K (1989) Nucleotide sequence of a full-length cDNA for 90 kDa heat-shock protein from human peripheral blood lymphocytes. Nucleic Acids Res 17: 7108–7108

Yang X-d, Gasser J, Riniker B, Feige U (1990) Treatment of adjuvant arthritis in rats: vaccination potential of a synthetic nonapeptide from the 65kDa heat shock protein of mycobacteria. J Autoimmun 3: 11–23

Young D, Lathigra R, Hendrix R, Sweetser D, Young RA (1988) Stress proteins are immune targets in leprosy and tuberculosis. Proc Natl Acad Sci USA 85: 4267–4270

Young DB, Ivanyi J, Cox JH, Lamb JR (1987) The 65 kDa antigen of mycobacteria—a common bacterial protein?. Immunol Today 8: 215–219

Young DB, Mehlert A, Bal V, Mendez Samperio P, Ivanyi J, Lamb JR (1988a) Stress proteins and the immune response to mycobacteria—antigens as virulence factors. Antonie Van Leeuwenhoek 54: 431–439

Young DB, Bal V, Mendez-Samperio P, Mehlert A, Ivanyi J, So A, Rothbard J, Jindal S, Young RA, Lamb JR (1988b) Stress proteins may provide a link between the immune response to infection and autoimmunity. Infect Immun

Young RA, Elliott TJ (1989) Stress proteins, infection, and immune surveillance. Cell 59: 5–8

Young RA, Mehra V, Sweetser D, Buchanan T, Clark Curtiss J, Davis RW, Bloom BR (1985) Genes for the major protein antigens of the leprosy parasite *Mycobacterium leprae*. Nature 316: 450–452

Yufu Y, Nishimura J, Takahira H, Ideguchi H, Nawata H (1989) Down-regulation of a Mr 90,000 heat shock cognate protein during granulocytic differentiation in HL-60 human leukemia cells. Cancer Res 49: 2405–2408

Zanetti M, Rogers J, Katz DH (1984) Induction of autoantibodies to thyroglobulin by anti-idiotypic antibodies. J Immunol 133: 240–243

Ziff M (1989) Pathways of mononuclear cell infiltration in rheumatoid synovitis. Rheumatol Int 9: 97–103

Ziff M, Cavender D, Haskard D (1988) Pathogenetic factors in rheumatoid synovitis. Br J Rheumatol 27: 153–156

Gamma/Delta T Lymphocytes and Heat Shock Proteins

S. H. E. Kaufmann[1] and D. Kabelitz[2]

1 Introduction

The importance of the immune system in protecting the host from exogenous invaders and endogenous aberrations is well appreciated. The underlying mechanisms are closely related to the unique capacity of the immune system to recognize foreign antigens and to distinguish them from self-antigens. T lymphocytes are central to this feature. Besides helping B cells in antibody production, T cells play a major role in defense against many infectious agents and in surveillance of tumor development. Aberrant T cell activation, on the other hand, can lead to autoimmune disease. T lymphocytes perform their role through two major pathways, namely first through the secretion of interleukins which activate a variety of effector cells and second, through the killing of target cells via direct cell contact. According to conventional thinking, T lymphocytes recognize antigenic peptides in the context of products of the major

[1] Department of Medical Microbiology and Immunology, University of Ulm, Albert-Einstein-Allee 11, 7900 Ulm, FRG
[2] Institute of Immunology, University of Heidelberg, Im Neuenheimer Feld 305, 6900 Heidelberg, FRG

Current Topics in Microbiology and Immunology, Vol. 167
© Springer-Verlag Berlin · Heidelberg 1991

histocompatibility gene complex (MHC). All T cells bear the CD3 marker on their surface; T cells expressing in addition the CD4 molecule normally recognize antigenic peptides in the context of MHC class II molecules, whereas CD8$^+$ T cells are normally specific for peptide plus MHC class I molecules.

Antigen is recognized by a disulfide-linked heterodimer termed T cell receptor (TCR). Until recently, it was assumed that only one TCR type composed of an α and a β chain exists (α/β TCR). We now know that a distinct second T cell lineage exists which expresses a γ/δ TCR (for review see BRENNER et al. 1988; RAULET 1989). Both the α/β and the γ/δ TCR are noncovalently linked with the CD3 complex which transduces the activation signal imparted by the antigen. The majority of γ/δ T cells lack both the CD4 and the CD8 molecule [i.e., they are double-negative (DN)], although some γ/δ T cells have been shown to express the CD8 or (to an even lesser extent) the CD4 molecule. Until recently, the antigen specificities, genetic restrictions, functional activities, and physiological role of these γ/δ T cells have been virtually unknown. Recent studies from several groups have identified mycobacterial components as frequent ligands for γ/δ T cells (JANIS et al. 1989; O'BRIEN et al. 1989; HOLOSHITZ et al. 1989; HAREGEWOIN et al. 1989; MODLIN et al. 1989; AUGUSTIN et al. 1989). In addition, evidence is emerging that heat shock proteins (hsp's) represent antigens for γ/δ T cells (O'BRIEN et al. 1989; HOLOSHITZ et al. 1989; HAREGEWOIN et al. 1989). In this treatise, the possible relation between hsp's and γ/δ T cells will be assessed. Evidence will be summarized to show that hsp's represent important antigens for certain γ/δ T cell sets. Evidence arguing against a common role of hsp's as antigens for γ/δ T cells in the periphery will be presented as well. Finally, possible types of interaction between hsp's and γ/δ T cells as well as potential biological consequences of this interaction will be discussed.

2 Heat Shock Proteins

Because hsp's are the central topic of this volume, only those characteristics of hsp's which are of particular relevance of γ/δ T cell recognition will be considered. The term "hsp" characterizes a group of proteins which are typically induced under heat shock (for review see KAUFMANN 1990). Some of these polypeptides are already present in normal cells and their synthesis is increased under shock; others are absent in normal cells and only appear at high temperature. According to their apparent molecular weight, different groups of hsp's can be distinguished (see Table 1). Importantly, hsp's are highly conserved with respect to both function and structure, and many of them seem to perform functions that are vital to normal cells—be they prokaryotic or eukaryotic. For example, hsp70 and hsp60 participate in the folding, unfolding, and translocation of polypeptides as well as in the assembly and disassembly of higher molecular complexes. The major function of hsp's after heat shock seems

Table 1. Physiological role of major hsp's

Family	Major cognates	Important function
hsp90	hsp90, hsp83	Interference with steroid receptor binding to DNA
hsp70	hsp70, BiP, hsc70, grp78, dnak	Protein folding, unfolding and translocation; assembly of multimeric complexes
hsp60	hsp65, groEL	Protein folding, unfolding, and translocation; assembly of multimeric complexes
Ubiquitin	Ubiquitin	Protein degradation

to be directly related to their capacity to reverse or to prevent the unfolding of polypeptides caused by increased temperature.

Not only heat, but many other assaults including anoxia, reactive oxygen metabolites, or depletion of essential ions induce hsp's. Insults of this type frequently occur in the host–parasite relationship. Inside professional phagocytes microorganisms are confronted with a variety of host defense mechanisms aimed at their eradication. Production of reactive oxygen metabolites represents an important intracellular killing mechanism and it can easily be envisaged that pathogens attempt to survive such attacks by means of elevated hsp levels. By increasing their hsp synthesis, they may become capable of resisting otherwise lethal stimuli.

H_2O_2, for example, causes hsp synthesis in *Salmonella typhimurium*, which subsequently become resistant against otherwise lethal doses of this molecule (CHRISTMAN et al. 1985; MORGAN et al. 1986). Moreover, *S. typhimurium* mutants with defective hsp genes are more readily killed by activated macrophages (FIELDS et al. 1986). On the other hand, infection also causes stress to the host cell. This has been well studied in the case of viral infections and more recent evidence suggests that similar rules hold true for bacterial infections, as well (for review see KAUFMANN 1990). For example, phagocytosis of particles as well as physiological stimulation with interleukins causes hsp synthesis in professional phagocytes (FERRIS et al. 1988; POLLA and KANTENGWA, this volume). Stress may also be caused by various other assaults a mammalian cell may have to face, like those occurring during inflammation, trauma, or transformation.

2.1 Heat Shock Proteins as T Cell Antigens

From the data discussed above it follows that microbes abundantly produce hsp's inside macrophages. As a corollary, hsp's can be processed by infected macrophages and presented on their surface in the context of MHC molecules. Therefore, hsp's should serve as T cell antigens of microbial pathogens. Indeed, data from several experimental systems are consistent with an immunodominant role of hsp's in the antimicrobial immune response. Thus, in mice immunized with *Mycobacterium tuberculosis*, about 20% of all T cells directed against mycobacterial organisms are specific for hsp60 (KAUFMANN et al. 1987). T cells against this hsp are also frequently found in leprosy and tuberculosis patients (for review see KAUFMANN et al. 1990).

In many healthy individuals T cells directed against hsp60 have also been identified and this finding argues against species specificity of the immune response against hsp's (MUNK et al. 1988). Because hsp's of different microbial organisms are highly homologous, it is possible that T cells against cross-reactive epitopes of hsp's are induced by a variety of microbes. Analysis of the fine specificity of T cells against hsp60 using synthetic peptides revealed that many T cells recognize epitopes that are identical with or highly similar to the hsp60 cognate in the host (MUNK et al. 1989; LAMB et al. 1989). T cells against such shared epitopes could be of potential harm for man since, by definition, they are autoreactive and hence could evoke autoimmune disease under certain conditions.

3 Gamma/Delta T Cells

3.1 Thymic Development

All T cells develop in the thymus. Thymic development of T lymphocytes proceeds via three major waves in ontogeny (for review see BRENNER et al. 1988; RAULET 1989). The first wave comprises DN γ/δ T cells, to a far lesser degree CD8$^+$CD4$^-$ γ/δ T cells and a minority of CD4$^+$CD8$^-$ γ/δ T cells; in the second wave, CD4$^+$CD8$^+$ α/β T cells, CD4$^+$CD8$^-$ α/β T cells, and CD8$^+$CD4$^-$ α/β T cells develop; in the third wave, DN α/β T cells arise. T cells of all three waves already express the CD3 molecular complex and are derived from precursors which are CD3$^-$CD4$^-$CD8$^-$. Rearrangement of the TCR γ and δ genes may already occur at the level of CD3$^-$CD4$^-$CD8$^-$ precursors directly leading to the appearance of CD3$^+$ γ/δ T cells.

In contrast to γ/δ T cells in the adult thymus, which are diverse, fetal thymic γ/δ T cells are highly homogeneous. In the mouse, around day 14 of gestation, a wave of γ/δ thymocytes appear which express particular $V_{\gamma5}$ and $V_{\delta1}$ chains (HAVRAN and ALLISON 1988), followed by a second wave comprising cells expressing particular $V_{\gamma6}$ and $V_{\delta1}$ gene combinations (ITO et al. 1989). $V_{\gamma5}$ and $V_{\gamma6}$ expression appears to be transient and they are not detected in the adult thymus. However, γ/δ T cells in certain epithelial tissue sites express the very same γ/δ TCR gene combinations and hence may directly originate from their thymic precursors (ASARNOW et al. 1988; ITOHARA et al. 1989; see below). [Unfortunately, at the moment, different nomenclatures are used for murine γ TCR genes, which may confuse the reader. This review adheres to the nomenclature of HEILIG and TONEGAWA (1987); cross-reference to designations by authors considered here is made in Table 2. More comprehensive comparisons are given by BRENNER et al. (1988) and RAULET (1989).]

Table 2. Correlation of two major nomenclatures for murine TCR V_γ genes and preferred V_γ localization or reactivity

V_γ designation according to (1)	V_γ designation according to (2)	Preferred localization or reactivity
1	1.1	Spleen, fetal, and adult thymus; PPD reactivity
2	1.2	Adult thymus
3	1.3	
4	2	Fetal thymus, PPD reactivity, lymphoid tissue
5	3	Fetal (day 14) thymus, skin
6	4	Fetal (day 18) thymus, uterus, vagina, tongue, PPD reactivity
7	5	Intestine

(1): HEILIG and TONEGAWA (1987);
(2): GARMAN et al. (1986); ASARNOW et al. (1988)

3.2 Distribution of γ/δ T Cells in the Periphery

In the peripheral blood of healthy humans, 1%–10% of the T lymphocytes express the γ/δ TCR; a similar percentage of γ/δ T cells is found in the secondary lymphoid organs of mice (for review see RAULET 1989). Interestingly, γ/δ T cells represent the major T cell type in the epithelia of skin, small intestine, reproductive organs, and probably also of the respiratory tract of mice. While most γ/δ T cells in the peripheral blood are DN, the majority of murine intestinal epithelial cells express the CD8 molecule. Epidermis, intestine, and respiratory tract provide a natural barrier to the environment. Skin and even more so intestine are inhabited by a plethora of microorganisms; although the lung is almost sterile, environmental microbes are constantly inhaled. Hence, these tissue sites must provide an important barrier for numerous pathogens and, when injured, serve as a major port of entry for many infections. It has, therefore, been proposed that γ/δ T cells perform an important surveillance role at these strategic locations (ASARNOW et al. 1988; BORN et al. 1990a; JANEWAY et al. 1988).

Murine γ/δ T cells present in these localizations preferentially use certain γ and δ TCR chain combinations (ASARNOW et al. 1988; BONNEVILLE et al. 1988; TAKAGAKI et al. 1989; ITOHARA et al. 1989, 1990). γ/δ cells of the intestine primarily express the $V_{\gamma7}$ chain paired with different V_δ chains and are structurally quite diverse (BONNEVILLE et al. 1988). γ/δ T cells in the skin, reproductive organs, and tongue are more homogeneous due to preferred usage of particular γ and δ chains and lack of junctional diversity (ASARNOW et al. 1988; ITOHARA et al. 1989). γ/δ T cells in the skin homogeneously express one particular $V_{\gamma5}$ and one particular $V_{\delta1}$ chain. The full $V_{\gamma5}J_{\gamma1}C_{\gamma1}$ combination used is also abundant in fetal (day 14) but not in adult thymocytes (see above). However, a $V_{\gamma1}$ expressing T cell line has also been isolated from murine epidermis (KONING et al. 1988), indicating that other γ chain usages are possible. The γ/δ T cells in the epithelia of vagina, uterus, and tongue preferentially express a canonical $V_{\gamma6}-V_{\delta1}$ sequence ($V_{\gamma6}J_{\gamma1}C_{\gamma1}$ and $V_{\delta1}D_\delta J_{\delta2}C_\delta$) which again is abundant in fetal (day 18) thymocytes. Such similarities in γ/δ TCR gene usage in fetal thymus and certain epithelia indicate direct development of the latter from the former. Furthermore, the restricted TCR diversity of γ/δ T cells in these epithelia implies recognition of a monomorphic ligand. This has led to the proposal that conserved molecules expressed on epithelial cells must serve as antigens (ASARNOW et al. 1988; JANEWAY et al. 1988; BORN et al. 1990a) and hsp's would be ideal candidates for γ/δ T cell recognition since they are ubiquitous, highly conserved, and indicative for diverse stress situations such as infection, inflammation, trauma, or transformation. Thus far, evidence for preferential localization of γ/δ T cells in epithelial tissue sites of humans is, however, less convincing (BUCY et al. 1989; GROH et al. 1989).

3.3 Genetic Restriction of γ/δ T Cells

Although evidence for antigen recognition in the context of MHC class I and MHC class II molecules by γ/δ T cells has been presented, several lines of data indicate that other restriction elements may be used, in addition (for review see STROMINGER 1989). In fact, it appears that γ/δ T cells recognize nonpolymorphic class Ib molecules (CD1, TL, and Qa) more often than polymorphic MHC class I or class II molecules.

3.4 Functional Activities of γ/δ T Cells

The first functional activity of γ/δ T cells to be described was their capacity to lyse a variety of targets in an MHC-independent fashion (for review see HERSEY and BOLHUIS 1987). Although, in the beginning, this observation facilitated the identification and characterization of γ/δ T cells, more recent evidence suggests that this so-called nonspecific killer activity can occur independent from the γ/δ TCR (PHILLIPS et al. 1987). Otherwise, γ/δ T lymphocytes express similar functional activities as do α/β T lymphocytes. Thus, specific killing via the γ/δ TCR has been described recently (RIVAS et al. 1989; MUNK et al. 1990). Various helper functions have also been associated with γ/δ T cells. γ/δ T cells which produce IL-2, IL-4, IFN-γ, GM-CSF, or less well defined interleukins after stimulation with mitogens, anti-CD3 mAb, or specific antigens have been described (FERRINI et al. 1987; BANK et al. 1986; BLUESTONE et al. 1989; MODLIN et al. 1989; KOZBOR et al. 1989; MUNK et al. 1990). At present, it remains unclear whether interleukin secretion and target cell lysis are functions of distinct subsets or whether they are mediated by a pluripotent cell type.

3.5 γ/δ T Cells and Mycobacteria

In 1989, several groups reported that mycobacteria act as strong stimulators of γ/δ T lymphocytes in both the human and the murine system. JANIS et al. (1989)

observed that immunization of mice with complete Freund's adjuvant results in a significant increase in γ/δ T cells which proliferate in response to mycobacterial lysates in vitro. O'BRIEN et al. (1989)—also in the murine system—established a battery of γ/δ T cell hybridomas from thymocytes of newborn mice. Many of these T cell hybridomas spontaneously produced IL-2. For several reasons it appeared likely that IL-2 production was stimulated through the γ/δ TCR. Interestingly IL-2 secretion by these spontaneously reactive γ/δ hybridomas was further increased by stimulation with PPD from *M. tuberculosis*. The vast majority of the PPD-reactive γ/δ T cell hybridomas expressed the $V_{\gamma1}J_{\gamma4}$ and $V_{\delta6}$ chain of the TCR (Happ et al. 1989). Other V_{γ} genes were not used while a few other V_{δ} gene products could be identified. In the mouse, γ/δ T lymphocytes seem to represent 8%–20% of resident pulmonary lymphocytes, and their number is further increased upon aerosol challenge with PPD from *M. tuberculosis*.

MODLIN et al. (1989) identified γ/δ T cells in skin lesions of leprosy patients suffering from reversal reactions and at the site of local lepromin (Mitsuda) reaction. γ/δ T lymphocytes isolated from these lesions were stimulated in vitro by mycobacterial antigens to produce factors that induce adhesion and aggregation of bone marrow-derived macrophages in the presence of GM-CSF. Evidence for accumulation of γ/δ T cells in and around the necrotic lesions of tuberculous lymphadenitis has also been described (FALINI et al. 1989).

γ/δ T cells isolated from the synovial fluid of rheumatoid arthritis patients recognize mycobacterial components in the apparent absence of HLA restriction (HOLOSHITZ et al. 1989), whereas a PPD-reactive γ/δ T cell line isolated from the peripheral blood of a tuberculin-positive healthy individual seemed to be genetically restricted (HAREGEWOIN et al. 1989). When peripheral blood leukocytes from normal healthy individuals were stimulated in vitro with mycobacterial lysates for 7–10 days, the relative percentage of γ/δ T cells increased 2 to > 20-fold (MUNK et al. 1990). In many individuals, bacteria other than *M. tuberculosis* induced selective expansion of γ/δ T cells, as well (e.g., group A streptococci, *Listeria monocytogenes*, or *M. leprae*). Stimulated γ/δ T cells appeared to be specific, because mycobacteria- or streptococci-activated γ/δ T cells only lysed targets primed with the homologous agent. In some cases, cross-reactive lysis was observed, indicating recognition of shared antigenic epitopes. Taken together, these data not only demonstrate a remarkable predilection of γ/δ T cells for mycobacteria but also raise the question of the relevance of this finding to protection against and pathogenesis of infections with mycobacteria and probably other microbial pathogens.

3.6 γ/δ T Cells and Heat Shock Proteins: Pro

Evidence that hsp's represent preferred antigens for γ/δ T cells among the multitude of mycobacterial components was provided by some of the studies discussed above. Thus, the PPD-reactive γ/δ T cell line from a normal individual described by HAREGEWOIN et al. (1989) (Fig. 1) as well as one γ/δ T cell clone from the synovial fluid of a rheumatoid arthritis patient tested by HOLOSHITZ et al.

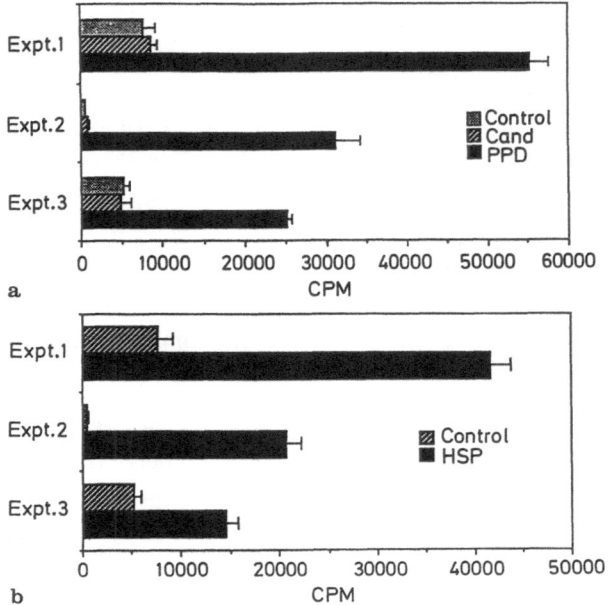

Fig. 1. Specific response of a γ/δ T cell line from a normal individual to PPD and hsp60. T cells were stimulated with PPD, candida or hsp60 and proliferative responses determined by ³H-TdR incorporation. (HAREGEWOIN et al. 1989)

(1989) responded to hsp60, as well. Similarly, O'BRIEN et al. (1989) found that many of their PPD-reactive γ/δ T cell hybridomas were also stimulated by the mycobacterial hsp60 (Fig. 2). More recently, the epitope recognized by these γ/δ T cell hybridomas was identified using a battery of overlapping synthetic peptides (BORN et al. 1990).

As mentioned above, subcutaneous or aerogenic immunization of mice with mycobacteria yields a high number of γ/δ T cells in lymph nodes or lung, respectively (JANIS et al. 1989; AUGUSTIN et al. 1989). More recent studies show that these γ/δ T cells can be further expanded in vitro by heat shock (RAJASEKAR et al. 1990). In this experiment, heat shock probably caused expression of the ligand seen by mycobacteria-activated γ/δ T cells in the accessory cells. Evidence for expression of cross-reactive epitopes shared by hsp's of mycobacteria and murine cells in the context of MHC class I gene products had already been given by KOGA et al. (1989). Pulmonary γ/δ T cells from mycobacteria-immune mice preferentially express $V_{\gamma 4}$ and $V_{\gamma 6}$ as well as $V_{\delta 4}$ and $V_{\delta 5}$. After heat shock, γ/δ T cells expressing $V_{\gamma 4}$ as well as $V_{\delta 5}$ and $V_{\delta 6}$ dominate with some $V_{\gamma 6}$ and $V_{\delta 4}$ cells remaining. These data indicate that pulmonary γ/δ T cells directed against shared epitopes of mycobacterial and mammalian hsp's preferentially use $V_{\gamma 4}$ with $V_{\delta 5}$ and $V_{\delta 6}$.

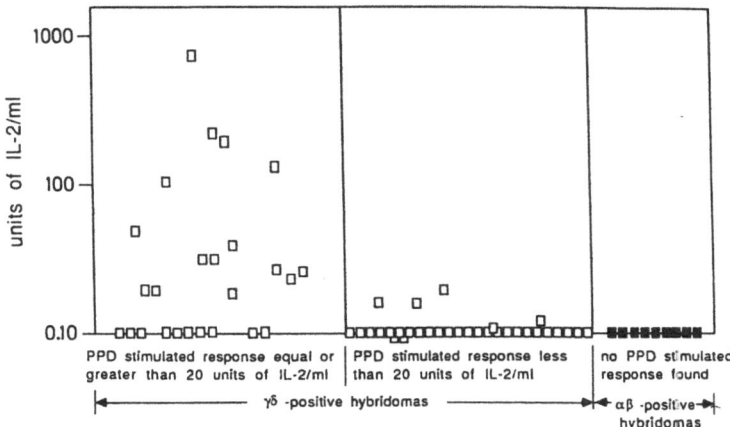

Fig. 2. Comparison of the reactivity to hsp60 and PPD by γ/δ T cell hybridomas derived from thymocytes. The figure shows IL-2 production after stimulation with hsp60 by γ/δ T cell hybridomas giving strong responses to PPD, by γ/δ T cell hybridomas giving medium responses to PPD, or by α/β T cell hybridomas not responsive to PPD. (O'BRIEN et al. 1989)

FISCH et al. (1990a) found that a major subset of γ/δ T cells which uses the $V_{\gamma 9}/V_{\delta 2}$ TCR genes mediates strong killing of Daudi cells but does not lyse Raji cells. This same subset preferentially recognizes mycobacterial lysates (FISCH et al. 1990b). Interestingly, a tentative association between hsp70 expression on target cells and their killing by γ/δ T cells expressing the $V_{\gamma 9}/V_{\delta 2}$ TCR was observed by these authors (FISCH et al. 1990b). Taken together, these findings, indeed, suggest that hsp's represent major antigens—at least for certain γ/δ T cells.

3.7 γ/δ T Cells and Heat Shock Proteins: Contra

Evidence arguing to the contrary has been obtained, as well. KABELITZ et al. (1990) isolated DN T cells which are highly enriched for γ/δ T cells (65%–93% in different experiments) and stimulated them with mycobacteria or with hsp60. In the presence of exogenous IL-2 unselected T cells as well as DN cells strongly proliferated in response to mycobacteria. In contrast, recombinant mycobacterial hsp60 was only weakly stimulatory. Limiting dilution analyses revealed an extremely high frequency of mycobacteria-responsive DN cells (Fig. 3). DN cells growing under limiting dilution conditions after stimulation with mycobacteria were more than 98% γ/δ^+ as assessed by FACS analysis. The frequency of DN cells which responded to recombinant hsp60 was much lower. In several experiments, 10 to 20-fold differences in frequencies between mycobacteria-reactive and hsp60-reactive DN cells were noted. In this study, several γ/δ T cell clones were established which responded well to mycobacteria

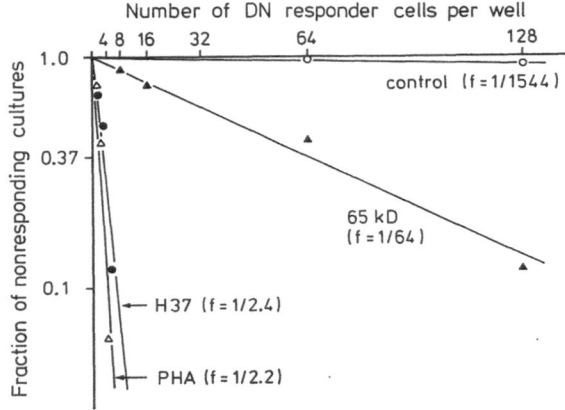

Fig. 3. Limiting dilution analysis of DN cells to *M. tuberculosis* or hsp60. DN cells (83% γ/δ T cells) were cultured under limiting dilution conditions with autologous feeder cells in the absence (control) or presence of phytohemagglutinin (PHA), *M. tuberculosis* (H37), or hsp60 (65 kD). After 14 days, proliferative responses were measured by ³H-TdR incorporation. Frequences (*f*) are given in the figure. (KABELITZ et al. 1990)

but not to mycobacterial hsp60. These findings argue against a major representation of hsp60-specific γ/δ T cells among the mycobacteria-reactive γ/δ T cell population present in the peripheral blood of healthy individuals.

PFEFFER et al. (1990) separated mycobacterial lysates according to their molecular weight and tested them on γ/δ T cells or on α/β T cells. α/β T cells primarily responded to fractions of 20- to 100-kD molecular mass, whereas γ/δ T cells preferentially reacted with fractions < 10 kD. Moreover, mycobacterial lysates, after extensive protease treatment, failed to stimulate α/β T cells but were still capable of stimulating γ/δ T cells. These observations indicate that a major fraction of mycobacteria-reactive γ/δ T cells recognize nonproteinaceous molecules and hence should not be hsp specific. Recognition of carbohydrates by γ/δ T cells has been proposed by STROMINGER (1989) and evidence can be found in the older literature indicating a role for carbohydrate antigens in protective immunity to tuberculosis (for review see CROWLE 1988). Alternatively, the slight possibility cannot be excluded that the protease-resistant material induced expression of a self-molecule which was ultimately recognized by γ/δ T cells. This molecule could be an hsp.

3.8 Possible Liaison Between γ/δ T Cells and Heat Shock Proteins

From the data discussed above it becomes clear that hsp's can serve as antigens for γ/δ T cells. However, they also show that hsp's are not as commonly seen by

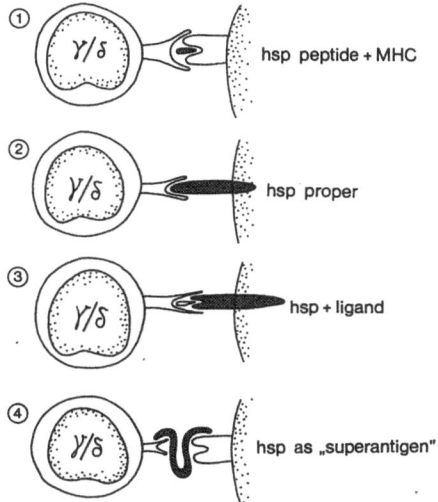

Fig. 4. Hypothetical scheme illustrating different types of hsp recognition by γ/δ T cells. *1*, γ/δ T cells recognize processed hsp's in the context of MHC molecules; *2*, γ/δ T cells recognize hsp's proper; *3*, γ/δ T cells recognize a ligand presented by an hsp; *4*, hsp's act as superantigen

γ/δ T cells as might have been anticipated in the beginning. Hence, a more critical evaluation of the relative importance of hsp-reactive γ/δ T cells is warranted. First, we shall assess how γ/δ T cells and hsp's could interact with each other and then we shall take a closer look at the functional capacities of hsp-reactive γ/δ T cells.

Evidence available thus far indicates different possibilities of contact between γ/δ T cells and hsp's which are not mutually exclusive (Fig. 4).

1. *γ/δ T cells recognize processed hsp's in the context of MHC molecules.* As discussed above, it is most likely that at least certain γ/δ T lymphocytes recognize hsp's in a conventional way, i.e., they interact with antigenic peptides derived from hsp's by classical processing in the context of MHC products. As restricting elements besides MHC class I and MHC class II products, class Ib molecules have to be considered in addition. Although the precise mechanisms of processing remain to be clarified, it appears likely that MHC class I association is preceded by cytosolic processing, whereas MHC class II association follows endosomal processing. Since self-hsp's are located in nonendosomal compartments, class I processing would be the most likely pathway for them. In contrast, bacterial hsp's should be preferentially processed through the MHC class II pathway. This possibility is supported by the findings of KOGA et al. (1989) indicating class I processing of hsp60 and those of BORN et al. (1990b) showing recognition of a synthetic peptide of hsp60 by murine γ/δ T cell hybridomas.

2. *γ/δ T cells recognize hsp's proper on the surface.* Although recognition by T lymphocytes of surface molecules in their pure form has not been convincingly demonstrated, this possibility cannot be excluded formally. Surface expression of hsp's has been described in several instances (KAUFMANN et al. 1990; VANBUSKIRK et al. 1989; KARLSSON-PARRA et al. 1990) and the observations of FISCH et al. (1990a, b) mentioned above would be consistent with such a type of interaction.

3. *γ/δ T cells recognize as yet undefined ligands that are presented by hsp's.* Several studies have already demonstrated a high binding affinity to hsp's to certain peptides (FLYNN et al. 1989), and VAN BUSKIRK et al. (1989) provided evidence for the involvement of hsp70 in the processing and presentation of a peptide to class II-restricted CD4 T cells.

4. *Hsp's act as superantigen.* Recent studies have revealed that enterotoxins stimulate certain T cell sets by linking the MHC class II molecule the accessory cells with given V_β chains of the TCR on the T cell (for review see FLEISCHER 1989). Evidence for *γ/δ* T cell stimulation by these so-called superantigens has been presented (FLEISCHER and SCHREZENMEIER 1988). Although the possibility that hsp's act as superantigens has not been formally excluded, it seems rather unlikely at the moment.

With respect to the functional scope of hsp-reactive *γ/δ* T cells, three major types of liaison can be envisaged (Fig. 5):

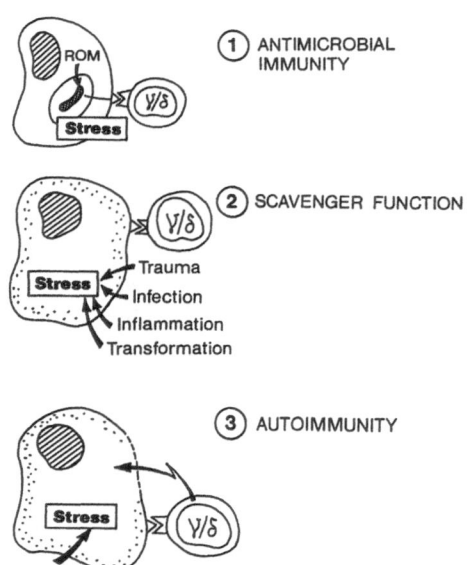

Fig. 5. Hypothetical scheme illustrating the functional potential of hsp-reactive *γ/δ* T cells. *1*, *γ/δ* T cells and antimicrobial immunity (ROM, reactive oxygen metabolites); *2*, *γ/δ* T cells and immune surveillance; *3*, *γ/δ* T cells and autoimmunity

1. γ/δ *T cells recognizing bacterial hsp's contribute to antimicrobial immunity.*
   During phylogeny hsp's have been highly conserved, and it has been shown,
   for example, that hsp's of various microorganisms share a remarkable degree
   of homology on the amino acid level (for review see KAUFMANN 1990). Hence,
   hsp's would be ideal markers for host cells which are infected by a wide variety
   of microorganisms. Because of their preferential localization in epithelial
   layers (at least in the mouse), γ/δ T cells would be ideally situated for scanning
   areas that are continuously exposed to microbial pathogens. Tubercle bacilli
   enter the host via the aerogenic route; hence the lung would be the first site of
   contact. Up to 20% of resident pulmonary $CD3^+$ lymphocytes express the γ/δ
   TCR (AUGUSTIN et al. 1989).

   Evidence has been presented for preferential $V_{\gamma 1}-V_{\delta 6}$ usage of thymic hsp-
   specific γ/δ T cells in the mouse (HAPP et al. 1989). More recent studies have
   revealed preferential usage of $V_{\gamma 4}$ and $V_{\gamma 6}$ with $V_{\delta 4}$ and $V_{\delta 5}$ in mycobacteria-
   reactive γ/δ T cells of mice from the lung and preferential usage of $V_{\gamma 4}$ with $V_{\delta 5}$
   and $V_{\delta 6}$ in those γ/δ T cells which react to heat-shocked cells (RAJASEKAR et al.
   1990). This usage does not correlate with that of γ/δ T cells in the epithelia of
   skin ($V_{\gamma 5}-V_{\delta 1}$), reproductive organs and tongue ($V_{\gamma 6}-V_{\delta 1}$), or intestine ($V_{\gamma 7}$).
   However, an epidermis-derived $CD3^+$ cell line expressing $V_{\gamma 1}$ was described by
   KONING et al. (1988).

   In experimental listeriosis of mice, the appearance of γ/δ T cells has been
   shown to precede that of α/β T cells (OHGA et al. 1990). Moreover, in vitro
   stimulation of lymph node cells from mycobacteria-immune mice results in a
   rapid burst of γ/δ T cells which is followed by preferential expansion of α/β T
   cells (JANIS et al. 1989). Finally, in leprosy lesions, γ/β T cells were only
   identified during reactional stages and in sites of delayed-type
   hypersensitivity reaction (MODLIN et al. 1989). Hence, the appearance of γ/δ T
   cells during bacterial infections seems to be restricted to certain stages of
   infection. Perhaps they are attracted to sites of inflammation after massive
   release or deposition of microbial antigens. Hsp-reactive γ/δ T cells could,
   therefore, provide for an early screening and scavenger mechanism against
   infectious agents of any species.

2. γ/δ *T cells recognizing host-derived hsp's contribute to surveillance of
   damaged or transformed cells.* γ/δ T cells which recognize host-derived hsp's
   could fulfil an even broader function and, besides infected host cells,
   recognize cells suffering from any insult such as inflammation, trauma,
   destruction, or transformation. HAREGEWOIN et al. (1990) recently showed that
   a γ/δ T cell clone from a healthy donor which recognizes mycobacterial hsp60
   also responds to the human hsp cognate. In a similar vein, the epitope
   recognized by the hsp-reactive mouse γ/δ T cell hybridomas described by
   O'BRIEN et al. (1989) shows about 50% sequence homology in the human
   cognate and the peptide homologue of the human hsp60 was capable of
   stimulating these γ/δ T cell hybridomas, though to a smaller degree (BORN
   et al. 1990b).

As a major prerequisite for recognition of "sick" host cells by γ/δ T cells they need to express their own hsp's on the surface (either alone or in the context of the appropriate restriction element, see above). Murine bone marrow-derived macrophages and B cells, as well as human synovial cells, are stained by monoclonal antibodies against hsp's (KAUFMANN et al. 1990; VANBUSKIRK 1989; KARLSSON-PARRA 1990). Furthermore, certain tumor cells seem to express cognates of the hsp70 and of the hsp90 family (FISCH et al. 1990b; SRIVASTAVA and MAKI, this volume). These findings indicate that host cells can express their own hsp's on the surface. Moreover, T lymphocytes activated against mycobacterial hsp60 were shown to lyse stressed bone marrow macrophages and Schwann cells in the absence of exogenous peptides (KOGA et al. 1989; STEINHOFF et al. 1990). Stress could be induced by IFN-γ stimulation, viral infection, and probably also bacterial infection. In contrast, unstimulated or uninfected macrophages and Schwann cells were not or only marginally lysed. Although the T cells used in these studies utilised the α/β TCR, the data strongly suggest that, under stress, autologous hsp60 is processed and presented in the context of MHC molecules and could be recognized by γ/δ T cells. Very recent evidence provides further support for recognition by γ/δ T cells of hsp epitopes shared by mycobacteria and stressed host cells (RAJASEKAR et al. 1990). The scavenger role of γ/δ T cells, therefore, may not be restricted to anti-infectious immunity, but may extend to a much broader scavenger system capable of recognizing stressed host cells independent of the type of assault.

3. *Hsp-reactive γ/δ T cells participate in autoimmune disease.* As long as the cells recognized by hsp-specific γ/δ T cells are somehow defective and hence destined to death, their elimination should be beneficial to the host. However, under certain conditions, the interaction of stressed host cells with γ/δ cells may induce a vicious circle which could ultimately cause autoimmune disease. The isolation of a γ/δ T cell clone with reactivity to hsp's from the synovial fluid of a rheumatoid arthritis patient may be taken as evidence for such a case (HOLOSHITZ et al. 1989).

4 Concluding Remarks

The idea that γ/δ T cells perform important scavenger functions, particularly in epithelial layers, is highly attractive. Although several arguments in favor of this assumption have been put forward, the puzzle is far from complete. Many more studies will be required before one can decide whether the liaison between hsp's and γ/δ T cells is indeed such a tight one or whether it is rather superficial.

Acknowledgments. S.K. gratefully acknowledges support by UNDP/World Bank/WHO Special Program for Research and Training in Tropical Diseases; Landesschwerpunkt 30; Sonderforschungsbereich 322; German Leprosy Relief Association; EEC-India Science and Technology Cooperation Program; A. Krupp award for young professors. Many thanks are due to R. Mahmoudi.

References

Asarnow DM, Kuziel WA, Bonyhadi M, Tigelaar RE, Tucker PW, Allison JP (1988) Limited diversity of $\gamma\delta$ antigen receptor genes of Thy-1$^+$ dendritic epidermal cells. Cell 55: 837–847

Augustin A, Kubo RT, Sim GK (1989) Resident pulmonary lymphocytes expressing the γ/δ T-cell receptor. Nature 340: 239–241

Bank I, DePinho RA, Brenner MB, Cassimeris J, Alt FW, Chess L (1986) A functional T3 molecule associated with a novel heterodimer on the surface of immature human thymocytes. Nature 322: 179–181

Bluestone J, Cron R, Cotterman M, Houlden B, Matis LA (1988) Structure and specificity of TCR γ/δ receptors on major histocompatibility complex antigen specific CD3$^+$, CD4$^-$, CD8$^-$ T lymphocytes. J Exp Med 168: 1899–1916

Bonneville M, Janeway CA Jr, Ito K, Haser W, Ishida I, Nakanishi N, Tonegawa S (1988) Intestinal intraepithelial lymphocytes are a distinct set of $\gamma\delta$ T cells. Nature 336: 479–481

Born W, Happ MP, Dallas A, Reardon C, Kubo R, Shinnick T, Brennan P, O'Brien R (1990a) Recognition of heat shock proteins and $\gamma\delta$ cell function. Immunol Today 11: 40–43

Born W, Hall L, Dallas A, Boymel J, Shinnick T, Young D, Brennan P, O'Brien R (1990b) Recognition of a peptide antigen by heat shock reactive $\gamma\delta$ T lymphocytes. Science 249: 67–69

Brenner MB, Strominger JL, Krangel MS (1988) The $\gamma\delta$ T cell receptor. Adv Immunol 43: 132–193

Bucy RP, Chen CLH, Cooper MD (1989) Tissue localization and CD8 accessory molecule expression of T$\gamma\delta$ cells in humans. J Immunol 142: 3045–3049

Christman MF, Morgan RW, Jacobson FS, Ames BN (1985) Positive control of a regulon for defenses against oxidative stress and some heat-shock proteins in Salmonella typhimurium. Cell 41: 753–762

Crowle AJ (1988) Immunization against tuberculosis. What kind of vaccine? Infect Immun 56: 2769–2773

Falini B, Flenghi L, Pileri S, Pelicci P, Fagiolo M, Martelli MF, Moretta L, Ciccone E (1989) Distribution of T cells bearing different forms of the T cell receptor γ/δ in normal and pathological human tissues. J Immunol 143: 2480–2488

Ferrini S, Bottino C, Blassoni R, Poggi A, Sekaly RP, Moretta L, Moretta L (1987) Characterization of CD3$^+$, CD4$^-$, CD8$^-$ clones expressing the putative T cell receptor δ gene product. J Exp Med 166: 277–282

Ferris DK, Harel-Bellan A, Morimoto RI, Welch WJ, Farrar WL (1988) Mitogen and lymphokine stimulation of heat shock proteins in T lymphocytes. Proc Natl Acad Sci USA 83: 5189–5193

Fields PI, Swanson RV, Haidaris CG, Heffrow F (1986) Mutants of Salmonella typhimurium that cannot survive within the macrophage are avirulent. Proc Natl Acad Sci USA 83: 5189–5193

Fisch P, Malkovsky M, Braakman E, Sturm E, Bolhuis RLH, Prieve A, Sosman JA, Lam VA, Sondel PM (1990a) T cell clones and natural killer cell clones mediate distinct patterns of non major histocompatibility restricted cytolysis. J Exp Med 171: 1567–1579

Fisch P, Malkovsky M, Klein BS, Welch WJ, Morrissey LW, Carper SW, Sondel PM (1990b) Human Vγ9/Vδ2 T-cells are stimulated by antigens from a Burkitt's lymphoma and mycobacteria. (submitted for publication)

Fleischer B (1989) Bacterial toxins as probes for the T-cell antigen receptor. Immunol Today 10: 262–264

Fleischer B, Schrezenmeier H (1988) T-cell stimulation by staphylococcal enterotoxins. Clonally variable response and requirement for major histocompatibility complex class II molecules on accessory or target cells. J Exp Med 167: 1697–1707

Flynn GC, Chappell TG, Rothman JE (1989) Peptide binding and release by proteins implicated as catalysts of protein assembly. Science 245: 385–390

Garman RD, Doherty PJ, Raulet DH (1986) Diversity, rearrangement and expression of murine T cell gamma genes. Cell 45: 733–742

Groh V, Porcelli S, Fabbi M, Lanier LL, Picker LJ, Anderson T, Warnke RA, Bhan AK, Strominger JL, Brenner MB (1989) Human lymphocytes bearing T cell receptor γ/δ are phenotypically diverse and evenly distributed throughout the lymphoid system. J Exp Med 169: 1277–1294

Happ MP, Kubo RT, Palmer E, Born WK, O'Brien R (1989) Limited receptor repertoire in a mycobacteria-reactive subset of $\gamma\delta$ T lymphocytes. Nature 342: 696–698

Haregewoin A, Soman G, Hom CR, Finberg RW (1989) Human $\gamma\delta^+$ T cells respond to mycobacterial heat-shock protein. Nature 340: 309–312

Haregewoin A, Singh B, Gupta RS, Finberg RW (1990) A mycobacterial heat shock protein responsive $\gamma\delta$ T cell clone also responds to the homologous human heat shock protein. J Inf Dis (in press)

Havran WL, Allison JP (1990) Origin of Thy-1$^+$ dendritic epidermal cells of adult mice from fetal thymic precursors. Nature 344: 68–70

Heilig JS, Tonegawa S (1987) T-cell γ gene is allelically but not isotypically excluded and is not required in known functional T-cell subsets. Proc Natl Acad Sci USA 84: 8070–8074

Hersey P, Bolhuis R (1987) "Nonspecific" MHC-unrestricted killer cells and their receptors. Immunol Today 8: 233–239

Holoshitz J, Koning F, Coligan JE, De Bruyn J, Strober S (1989) Isolation of CD4$^-$ CD8$^-$ mycobacteria-reactive T lymphocyte clones from rheumatoid arthritis synovial fluid. Nature 339: 226–229

Ito K, Bonneville M, Takagaki Y, Nakanishi N, Kanagawa O, Krecko EG, Tonegawa S (1989) Different γ/δ T-cell receptors are expressed on thymocytes at different stages of development. Proc Natl Acad Sci USA 86: 631–635

Itohara S, Nakanishi N, Kanagawa O, Kubo R, Tonegawa S (1989) Monoclonal antibodies specific to native murine T-cell receptor $\gamma\delta$: Analysis of $\gamma\delta$ T cells during thymic ontogeny and in peripheral lymphoid organs. Proc Natl Acad Sci USA 86: 5094–5098

Itohara S, Farr AG, Lafaille JJ, Bonneville M, Takagaki Y, Haas W, Tonegawa S (1990) Homing of a $\gamma\delta$ thymocyte subset with homogeneous T-cell receptors to mucosal epithelia. Nature 343: 754–757

Janeway CA Jr, Jones B, Hayday A (1988) Specificity and function of T cells bearing $\gamma\delta$ receptors. Immunol Today 9: 73–76

Janis EM, Kaufmann SHE, Schwartz RH, Pardoll DM (1989) Activation of $\gamma\delta$ T cells in the primary immune response to *Mycobacterium tuberculosis*. Science 244: 713–716

Kabelitz D, Bender A, Schondelmaier S, Schoel B, Kaufmann SHE (1990) A large fraction of human peripheral blood γ/δ^+ T cells is activated by *Mycobacterium tuberculosis* but not by its 65-kD heat shock protein. J Exp Med 171: 667–679

Karlsson-Parra A, Söderström K, Ferm M, Ivanyi J, Kiessling R, Klareskog L (1990) Presence of human 65 kD heat shock protein (hsp) in inflamed joints and subcutaneous nodules of RA patients. Scand J Immunol 31: 283–288

Kaufmann SHE (1990) Heat shock proteins and the immune response. Immunol Today 11: 129–136

Kaufmann SHE, Väth U, Thole JER, van Embden JDA, Emmrich F (1987) Enumeration of T cells reactive with *Mycobacterium tuberculosis* organisms and specific for the recombinant mycobacterial 64 kiloDalton protein. Eur J Immunol 17: 351–357

Kaufmann SHE, Schoel B, Wand-Württenberger A, Steinhoff U, Munk ME, Koga T (1990) T-cells, stress proteins, and pathogeneis of mycobacterial infections. Curr Top Microbiol Immunol 155: 125–141

Koga T, Wand-Württenberger A, DeBruyn J, Munk ME, Schoel B, Kaufmann SHE (1989) T cells against a bacterial heat shock protein recognize stressed macrophages. Science 245: 1112–1115

Koning F, Yokoyama WM, Maloy WL, Stingl G, McConnell TJ, Cohen DI, Shevach EM, Coligan JE (1988) Expression of Cγ4 T cell receptors and lack of isotype exclusion by dendritic epidermal T cell lines. J Immunol 141: 2057–2062

Kozbor D, Trinchieri G, Monos DS, Isobe M, Russo G, Haney JA, Zmijewski C, Croce CM (1989) Human TCR-γ^+/δ^+, CD8$^+$ T lymphocytes recognize tetanus toxoid in an MHC-restricted fashion. J Exp Med 169: 1847–1851

Lamb JR, Bal V, Mendez-Samperio P, Mehlert A, So A, Rothbard J, Jindal S, Young RA, Young DB (1989) Stress proteins may provide a link between the immune response to infection and autoimmunity. Int Immunol 1: 191–196

Modlin RL, Pirmez C, Hofmann FM, Torigian V, Uyemura K, Rea TH, Bloom BR, Brenner MB (1989) Lymphocytes bearing antigen-specific $\gamma\delta$ T-cell receptors accumulate in human infectious disease lesions. Nature 339: 544–548

Morgan RW, Christman MF, Jacobson FS, Storz G, Ames BN (1986) Hydrogen peroxide-inducible proteins in *Salmonella typhimurium* overlap with heat shock and other stress proteins. Proc Natl Acad Sci USA 83: 8059–8063

Munk ME, Schoel B, Kaufmann SHE (1988) T cell responses of normal individuals towards recombinant protein antigens of *Mycobacterium tuberculosis*. Eur J Immunol 18: 1835–1838

Munk ME, Schoel B, Modrow S, Karr RW, Young RA, Kaufmann SHE (1989) Cytolytic CD4$^+$ T lymphocytes from healthy individuals with specificity to self epitopes shared by the mycobacterial and human 65 kDa heat shock protein. J Immunol 143: 2844–2849

Munk ME, Gatrill AJ, Kaufmann SHE (1990) Antigen-specific target cell lysis and interleukin-2 secretion by *Mycobacterium tuberculosis* activated γ/δ T lymphocytes. (submitted for publication)

O'Brien R, Happ MP, Dallas A, Palmer E, Kubo R (1989) Stimulation of a major subset of lymphocytes expressing T cell receptor δ by an antigen derived from *Mycobacterium tuberculosis*. Cell 57: 667–674

Ohga S, Yoshikai Y, Takeda Y, Hiromatsu K, Nomoto K (1990) Sequential appearance of γ/δ- and α/β-bearing T cells in the peritoneal cavity during an i.p. infection with *Listeria monocytogenes*. Eur J Immunol 20: 533–538

Pfeffer K, Schoel B, Gulle H, Kaufmann SHE, Wagner H (1990) Primary responses of human T cells to mycobacteria: a frequent set of γ/δ T cells are stimulated by protease-resistant ligands. Eur J Immunol 20: 1175–1179

Phillips JH, Weiss A, Gemlo BT, Rayner AA, Lanier LL (1987) Evidence that the T cell antigen receptor may not be involved in cytotoxicity mediated by γ/δ and α/β thymic cell lines. J Exp Med 166: 1579–1584

Rajasekar R, Sim GK, Augustin A (1990) Self heat shock and $\gamma\delta$ T-cell reactivity. Proc Natl Acad Sci USA 87: 1767–1771

Raulet DH (1989) The structure, function, and molecular genetics of the γ/δ T cell receptor. Ann Rev Immunol 7: 175–207

Rivas A, Koide J, Cleary ML, Engleman EG (1989) Evidence for involvement of the γ, δ T cell antigen receptor in cytotoxicity mediated by human alloantigen-specific T cell clones. J Immunol 142: 1840–1846

Steinhoff U, Schoel B, Kaufmann SHE (1990) Lysis of interferon-γ activated Schwann cells by crossreactive CD8 α/β T cells with specificity to the mycobacterial 65 kDa heat shock protein. Int Immunol 2: 279–284

Strominger JL (1989) The $\gamma\delta$ T cell receptor and class Ib MHC-related proteins: enigmatic molecules of immune recognition. Cell 57: 895–898

Takagaki Y, DeCloux A, Bonneville M, Tonegawa S (1989) Diversity of $\gamma\delta$ T-cell receptors on murine intestinal intraepithelial lymphocytes. Nature 339: 712–714

VanBuskirk A, Crump BL, Margoliash E, Pierce SK (1989) A peptide binding protein having a role in antigen presentation is a member of the HSP70 heat shock family. J Exp Med 170: 1799–1809

Subject Index

Current Topics in Microbiology and Immunology

Volumes published since 1985 (and still available)

Vol. 115: **Vogt, Peter K. (Ed.):** Human T-Cell Leukemia Virus. 1985. 74 figs. IX, 266 pp. ISBN 3-540-13963-X

Vol. 116: **Willis, Dawn B. (Ed.):** Iridoviridae. 1985. 65 figs. X, 173 pp. ISBN 3-540-15172-9

Vol. 122: **Potter, Michael (Ed.):** The BALB/c Mouse. Genetics and Immunology. 1985. 85 figs. XVI, 254 pp. ISBN 3-540-15834-0

Vol. 124: **Briles, David E. (Ed.):** Genetic Control of the Susceptibility to Bacterial Infection. 1986. 19 figs. XII, 175 pp. ISBN 3-540-16238-0

Vol. 125: **Wu, Henry C.; Tai, Phang C. (Ed.):** Protein Secretion and Export in Bacteria. 1986. 34 figs. X, 211 pp. ISBN 3-540-16593-2

Vol. 126: **Fleischer, Bernhard; Reimann, Jörg; Wagner, Hermann (Ed.):** Specificity and Function of Clonally Developing T-Cells. 1986. 60 figs. XV, 316 pp. ISBN 3-540-16501-0

Vol. 127: **Potter, Michael; Nadeau, Joseph H.; Cancro, Michael P. (Ed.):** The Wild Mouse in Immunology. 1986. 119 figs. XVI, 395 pp. ISBN 3-540-16657-2

Vol. 128: 1986. 12 figs. VII, 122 pp. ISBN 3-540-16621-1

Vol. 129: 1986. 43 figs., VII, 215 pp. ISBN 3-540-16834-6

Vol. 130: **Koprowski, Hilary; Melchers, Fritz (Ed.):** Peptides as Immunogens. 1986. 21 figs. X, 86 pp. ISBN 3-540-16892-3

Vol. 131: **Doerfler, Walter; Böhm, Petra (Ed.):** The Molecular Biology of Baculoviruses. 1986. 44 figs. VIII, 169 pp. ISBN 3-540-17073-1

Vol. 132: **Melchers, Fritz; Potter, Michael (Ed.):** Mechanisms in B-Cell Neoplasia. Workshop at the National Cancer, Institute, National Institutes of Health, Bethesda, MD, USA, March 24–26, 1986. 1986. 156 figs. XII, 374 pp. ISBN 3-540-17048-0

Vol. 133: **Oldstone, Michael B. (Ed.):** Arenaviruses. Genes, Proteins, and Expression. 1987. 39 figs. VII, 116 pp. ISBN 3-540-17246-7

Vol. 134: **Oldstone, Michael B. (Ed.):** Arenaviruses. Biology and Immunotherapy. 1987. 33 figs. VII, 242 pp. ISBN 3-540-17322-6

Vol. 135: **Paige, Christopher J.; Gisler, Roland H. (Ed.):** Differentiation of B Lymphocytes. 1987. 25 figs. IX, 150 pp. ISBN 3-540-17470-2

Vol. 136: **Hobom, Gerd; Rott, Rudolf (Ed.):** The Molecular Biology of Bacterial Virus Systems. 1988. 20 figs. VII, 90 pp. ISBN 3-540-18513-5

Vol. 137: **Mock, Beverly; Potter, Michael (Ed.):** Genetics of Immunological Diseases. 1988. 88 figs. XI, 335 pp. ISBN 3-540-19253-0

Vol. 138: **Goebel, Werner (Ed.):** Intracellular Bacteria. 1988. 18 figs. IX, 179 pp. ISBN 3-540-50001-4

Vol. 139: **Clarke, Adrienne E.; Wilson, Ian A. (Ed.):** Carbohydrate-Protein Interaction. 1988. 35 figs. IX, 152 pp. ISBN 3-540-19378-2

Vol. 140: **Podack, Eckhard R. (Ed.):** Cytotoxic Effector Mechanisms. 1989. 24 figs. VIII, 126 pp. ISBN 3-540-50057-X

Vol. 141: **Potter, Michael; Melchers, Fritz (Ed.):** Mechanisms in B-Cell Neoplasia 1988. Workshop at the National Cancer Institute, National Institutes of Health, Bethesda, MD, USA, March 23–25, 1988. 1988. 122 figs. XIV, 340 pp. ISBN 3-540-50212-2

Vol. 142: **Schüpach, Jörg:** Human Retrovirology. Facts and Concepts. 1989. 24 figs. 115 pp. ISBN 3-540-50455-9

Vol. 143: **Haase, Ashley T.; Oldstone, Michael B. A. (Ed.):** In Situ Hybridization. 1989. 33 figs. XII, 90 pp. ISBN 3-540-50761-2

Vol. 144: **Knippers, Rolf; Levine, A. J. (Ed.):** Transforming Proteins of DNA Tumor Viruses. 1989. 85 figs. XIV, 300 pp. ISBN 3-540-50909-7

Vol. 145: **Oldstone, Michael B. A. (Ed.):** Molecular Mimicry. Cross-Reactivity between Microbes and Host Proteins as a Cause of Autoimmunity. 1989. 28 figs. VII, 141 pp. ISBN 3-540-50929-1

Vol. 146: **Mestecky, Jiri; McGhee, Jerry (Ed.):** New Strategies for Oral Immunization. International Symposium at the University of Alabama at Birmingham and Molecular Engineering Associates, Inc. Birmingham, AL, USA, March 21–22, 1988. 1989. 22 figs. IX, 237 pp. ISBN 3-540-50841-4

Vol. 147: **Vogt, Peter K. (Ed.):** Oncogenes. Selected Reviews. 1989. 8 figs. VII, 172 pp. ISBN 3-540-51050-8

Vol. 148: **Vogt, Peter K. (Ed.):** Oncogenes and Retroviruses. Selected Reviews. 1989. XII, 134 pp. ISBN 3-540-51051-6

Vol. 149: **Shen-Ong, Grace L. C.; Potter, Michael; Copeland, Neal G. (Ed.):** Mechanisms in Myeloid Tumorigenesis. Workshop at the National Cancer Institute, National Institutes of Health, Bethesda, MD, USA, March 22, 1988. 1989. 42 figs. X, 172 pp. ISBN 3-540-50968-2

Vol. 150: **Jann, Klaus; Jann, Barbara (Ed.):** Bacterial Capsules. 1989. 33 figs. XII, 176 pp. ISBN 3-540-51049-4

Vol. 151: **Jann, Klaus; Jann, Barbara (Ed.):** Bacterial Adhesins. 1990. 23 figs. XII, 192 pp. ISBN 3-540-51052-4

Vol. 152: **Bosma, Melvin J.; Phillips, Robert A.; Schuler, Walter (Ed.):** The Scid Mouse. Characterization and Potential Uses. EMBO Workshop held at the Basel Institute for Immunology, Basel, Switzerland, February 20–22, 1989. 1989. 72 figs. XII, 263 pp. ISBN 3-540-51512-7

Vol. 153: **Lambris, John D. (Ed.):** The Third Component of Complement. Chemistry and Biology. 1989. 38 figs. X, 251 pp. ISBN 3-540-51513-5

Vol. 154: **McDougall, James K. (Ed.):** Cytomegaloviruses. 1990. 58 figs. IX, 286 pp. ISBN 3-540-51514-3

Vol. 155: **Kaufmann, Stefan H. E. (Ed.):** T-Cell Paradigms in Parasitic and Bacterial Infections. 1990. 24 figs. IX, 162 pp. ISBN 3-540-51515-1

Vol. 156: **Dyrberg, Thomas (Ed.):** The Role of Viruses and the Immune System in Diabetes Mellitus. 1990. 15 figs. XI, 142 pp. ISBN 3-540-51918-1

Vol. 157: **Swanstrom, Ronald; Vogt, Peter K. (Ed.):** Retroviruses. Strategies of Replication. 1990. 40 figs. XII, 260 pp. ISBN 3-540-51895-9

Vol. 158: **Muzyczka, Nicholas (Ed.):** Viral Expression Vectors. 1990. approx. 20 figs. approx. XII, 190 pp. ISBN 3-540-52431-2

Vol. 159: **Gray, David; Sprent, Jonathan (Ed.):** Immunological Memory. 1990. 38 figs. XII, 156 pp. ISBN 3-540-51921-1

Vol. 160: **Oldstone, Michael B. A.; Koprowski, Hilary (Ed.):** Retrovirus Infections of the Nervous System. 1990. 16 figs. XII, 176 pp. ISBN 3-540-51939-4

Vol. 161: **Racaniello, Vincent R. (Ed.):** Picornaviruses. 1990. 12 figs. X, 194 pp. ISBN 3-540-52429-0

Vol. 162: **Roy, Polly; Gorman, Barry M. (Ed.):** Bluetongue Viruses. 1990. 37 figs. X, 200 pp. ISBN 3-540-51922-X

Vol. 163: **Turner, Peter C.; Moyer, Richard W. (Ed.):** Poxviruses. 1990. 23 figs. X, 210 pp. ISBN 3-540-52430-4

Vol. 164: **Bækkeskov, Steinnun; Hansen, Bruno (Ed.):** Human Diabetes. 1990. 9 figs. X, 198 pp. ISBN 3-540-52652-8

Vol. 165: **Bothwell, Mark (Ed.):** Neuronal Growth Factors. 1990. 14 figs. approx. X, 200 pp. ISBN 3-540-52654-4

Vol. 166: **Potter, Michael; Melchers, Fritz (Ed.):** Mechanisms in B-Cell Neoplasia 1990. 1990. 143 figs. XIX, 380 pp. ISBN 3-540-52886-5